From Colonization to Nation-State

Riwanto Tirtosudarmo

From Colonization to Nation-State

The Political Demography of Indonesia

Springer

LIPI

Riwanto Tirtosudarmo
Indonesian Institute of Sciences
Jakarta, Indonesia

ISBN 978-981-16-6436-6 ISBN 978-981-16-6437-3 (eBook)
https://doi.org/10.1007/978-981-16-6437-3

Jointly published with Indonesian Institute of Sciences (LIPI) Press.
LIPI Press retains the right to market and sell the print version of the Work in English language in Indonesia.

Originally published by Indonesian Institute of Sciences (LIPI), Jakarta, 2013
1th edition: © Indonesian Institute of Sciences (LIPI) 2013

This Springer imprint is published by the registered company Springer Nature Singapore Pte Ltd.
The registered company address is: 152 Beach Road, #21-01/04 Gateway East, Singapore 189721, Singapore

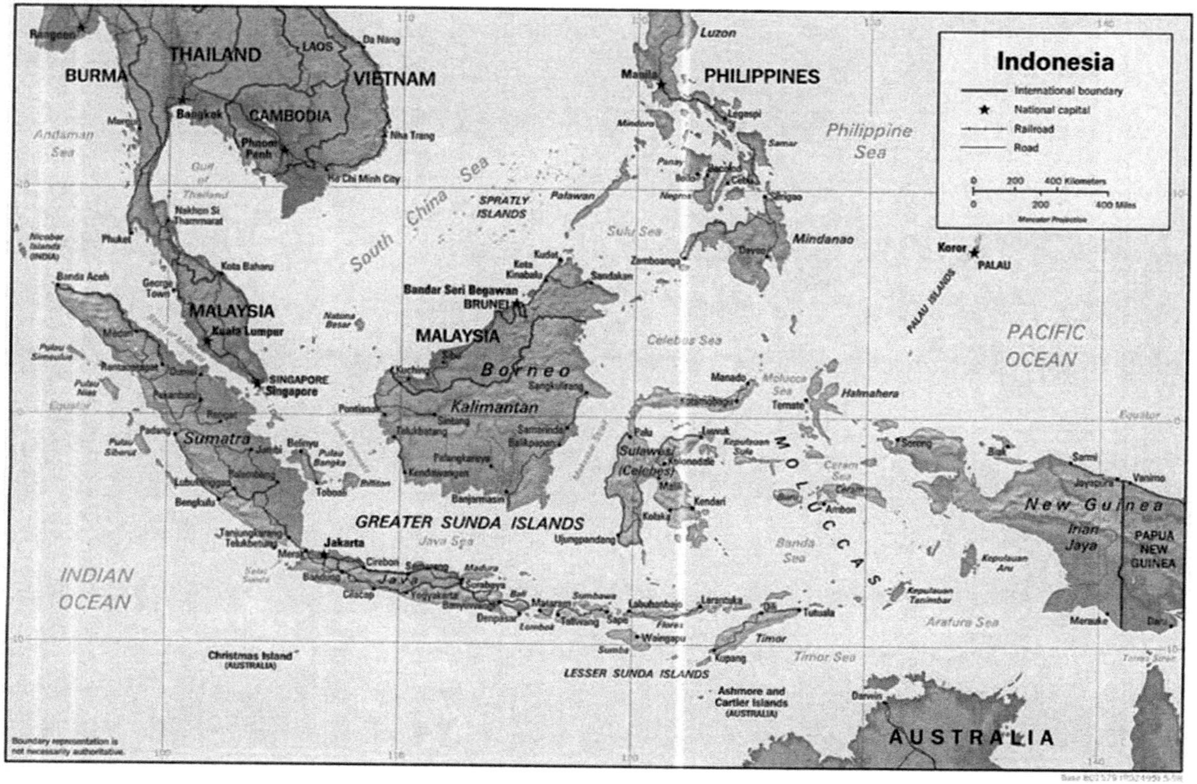
Indonesia
International boundary
National capital
Railroad
Road
0 200 400 Kilometers
0 200 400 Miles
Mercator Projection
BURMA
THAILAND
LAOS
VIETNAM
CAMBODIA
MALAYSIA
SINGAPORE
BRUNEI
PHILIPPINES
PALAU
PAPUA NEW GUINEA
AUSTRALIA
Rangoon
Bangkok
Phnom Penh
Da Nang
Nha Trang
Ho Chi Minh City
Kuala Lumpur
Singapore
Bandar Seri Begawan
Manila
Koror
Andaman Sea
Gulf of Thailand
South China Sea
Sulu Sea
Celebes Sea
Molucca Sea
Philippine Sea
PACIFIC OCEAN
INDIAN OCEAN
Java Sea
Banda Sea
Ceram Sea
Arafura Sea
Timor Sea
SPRATLY ISLANDS
PALAU ISLANDS
Luzon
Mindoro
Palawan
Panay
Negros
Samar
Mindanao
Legaspi
Zamboanga
Davao
Sumatra
Borneo
Kalimantan
Sulawesi (Celebes)
Java
Halmahera
New Guinea
Irian Jaya
Timor
Flores
Sumba
Bali
Lombok
Madura
Sumbawa
MOLUCCAS
GREATER SUNDA ISLANDS
LESSER SUNDA ISLANDS
Banda Aceh
Medan
Padang
Pekanbaru
Jambi
Palembang
Bengkulu
Jakarta
Bandung
Cirebon
Semarang
Surabaya
Yogyakarta
Cilacap
Denpasar
Kupang
Pontianak
Banjarmasin
Balikpapan
Samarinda
Palangkaraya
Kuching
Sibu
Kota Kinabalu
Kudat
Sandakan
Manado
Ternate
Ujungpandang
Kendari
Palu
Ambon
Sorong
Jayapura
Merauke
Vanimo
Darwin
Christmas Island (AUSTRALIA)
Ashmore and Cartier Islands (AUSTRALIA)
Nicobar Islands (INDIA)
Equator
Boundary representation is not necessarily authoritative.

Foreword

Even though the pandemic of COVID-19 has temporarily hampered people's move from one place to another, migration never ceases. Human beings, from the very beginning, have migrated, and the move has been accelerated in parallel with the waves of globalization. Migration, being an essential factor of human society, was, however, treated as an additional matter by a still popular notion of nation-state which consists of people, society, culture, and boundaries. Migration studies, on the other hand, have sometimes remained in the economic and the demographic domain, and not been developed into a wider and richer field.

Seeing the main title of this book, readers may have an impression that the author discusses decolonization process of Indonesia. The author does discuss colonization and building of a nation-state. However, he does so not by describing merely the process of colonization, independence, and nation-building. Instead, Dr. Riwanto Tirtosudarmo, a renowned social scientist of Indonesia, by focusing on transmigration in Indonesia, demonstrates the ways of discussing the political factors and agents behind the transmigration and consequences of transmigration from the viewpoint of "political demography." Readers will realize how the domestic transmigration programs and the government's attitude towards international migration are linked, in positive as well as negative ways, with Indonesian nation-building that is yet to be completed. Dr. Riwanto, an expert of social demography, has stepped into a new dimension which encompasses migration, local societies, and politics, domestic as well as international.

As Dr. Riwanto rightly points out, migration is a multifaceted phenomenon in which one will be able to observe domestic politics, economic benefit, international politics; manipulation of ethnic images; social problems both among hosts and guests; and business-related migration.

Dr. Riwanto presents a variety of examples from Indonesia, the country characterized by persistent surplus and imbalance of labor. While he spares many pages on the situation in Kalimantan, the topics he deals extend from ethnic conflicts to environmental degradation, from colonial to modern era, and from East Timor to

West Irian. Nevertheless, behind the issues, Dr. Riwanto consistently sees the intents, omission, incongruence, and indifference in the sphere of politics.

The chapters in this book demonstrate Indonesian situation seen from a migration perspective and at the same time suggest a new approach to migration studies by stressing political factors.

Emeritus Professor of Anthropology, Koji Miyazaki
Tokyo University of Foreign Studies,
Tokyo, Japan

Publisher's Note

As a scientific publisher, BRIN Publishing (formerly known as LIPI Press) holds high responsibility to enlighten society intelligence and awareness through the provision of qualified publication. This scientific book entitled *From Colonization to Nation-State: the Political Demography of Indonesia* was once published by LIPI Press in 2013. This book has encountered quality control mechanism through the editorial process, including peer review and copy editing.

The book critically assesses the continuities and changes in the state's demographic engineering practices from the introduction of the Dutch colonial ethical policy at the beginning of the twentieth century until the dawn of the twenty-first century following the collapse of Indonesia's New Order regime in 1998. Several issues that are covered in the book include transmigration policy, internal and cross-border mobility, ethnic mobility, and the politics of migration.

We surely hope this book could give new insights and information, especially about the dynamic nature of demography and its long-term profound implication for social and political changes. The book is also expected to contribute to a better understanding of Indonesia's complex social and political changes, which is currently approaching a new historical crossroad of the twenty-first century. With the increasing interconnectedness of countries through flows and movements of people, the influence of Indonesia's political, demographic changes will not only be within but will eventually be further beyond its border.

As a final note, we would like to express our heartfelt gratitude to everyone taking part in the process of this book.

BRIN Publishing

Introduction: The Political Demography of Nation-State Building

As one might expect in a region with deeper sources of political instability and fewer democratic traditions, the ways in which population issues and politics have intersected have been harsher.

Michael S. Teitelbaum and Jay Winter (1998, 5)

At the dawn of the twenty-first century, Indonesia is entering a fresh chapter in its history, when a new narrative must be written on the fate of this archipelago. Borrowing the phraseology of two important collections of essays, which represent the scholarly mood of the day: "the nation-state is at bay or under siege,"[1] perhaps, we should also ponder on the future of Indonesia, at this historical trajectory. Prior to, and in the aftermath of Suharto's New Order step down from power in May 1998, a series of events indicates that Indonesia was racked by many incidents of social and political conflict that put the idea of Indonesia as a nation-state "at bay" or "under siege." The rapid, social, and political transformation currently occurring eventually could change the reconfiguration of the existing nation-state formation. The most critical phenomenon in this galloping makeover is the increasing political mobilization, based on seemingly primordial sentiments related to territory, ethnicity, and religion. Ethnicity, in a broader sense, seems a common platform, often subtly interwoven with the notion of territory and religion, transforming the current political organization and making it complicated and difficult to untangle. The central position of ethnicity, in various forms of social and political conflicts, puts Indonesia, perhaps, in a comparable situation to that experienced, previously, in the Eastern European countries and in the Soviet Union, in the aftermath of the Cold War. Yet, as Maybury-Lewis (1997, ix) notes:

Ethnicity is one of the obscurest issues of them all. In spite of the fact that there is much talk nowadays about ethnicity and ethnic groups, it is not always clear exactly what that means or why ethnicity should have such a powerful hold on human imagination and behavior.

---

[1] Young, C. (ed.). 1993. *The Rising Tide of Cultural Pluralism: The Nation State at Bay?* and Ferguson, R. B. (ed.). 2004. *The State Under Siege: Political Disintegration in the Post-Cold War Era.*

> *Ethnic conflict has supplanted communism as the specter that is haunting the world. Writers, commentators and newspapers are currently telling us that people are everywhere giving in to the primordial urge to band together with others like themselves and to harass or kill those who are different.*

What is written by Marybury-Lewis in 1997, more than two decades ago, has been continuously ringing in our ears as within the last two decades indeed ethnicity that is then manifested in the so-called politics of identity and the resurgence of populism transforms the world into the current situation. In Indonesia, the postcolonial discourse on ethnicity has been developing in tandem with the increasing state domination of society. Although the national motto "Unity in Diversity" is constantly propagated, the state practice recurrently emphasizes the importance of cultural uniformity and disregards ethnic plurality. The idea of national integration, an obsession of the ruling elites, has strongly influenced the construction of differences and heterogeneity as a perceived national threat. Such a threat, again in the perception of the ruling elites, has to be eliminated in order to achieve the imagined national unity. Assimilation and harmony, two keywords translated into state policies and programs, must be designed to realize national integration and economic prosperity. The omission of questions on ethnicity in population censuses, after Independence, perhaps, is also deeply rooted in the ruling elites' perception of the notion of a national threat emanating from any demographic exposure of ethnic identity. Exposition of such heterogeneity, perceived as possibly enhancing *sukuisme* (ethnocentrism, primordial sentiment), and in turn, threatening the realization of Indonesian national integration. The Central Bureau of Statistics, as the state institution responsible for conducting population censuses, is constantly under pressure from the higher state officials to erase any questions on ethnicity.[2] Population politics, in which demographic change is regarded as making a significant contribution to politics, has apparently been inherent within the national elite psyche since the beginning of independence. Demographic engineering, as a mechanism for obtaining the desired cultural homogeneity and ultimately, national integration, is an

[2]This information was obtained by the author from a conversation in September 2000 with one of the senior staff of the Central Bureau of Statistics who has been involved in the preparation of population censuses including playing an important role in the preparation of the 2000 Population Census. Statistical information on ethnicity is therefore lacking; if there is any, it is usually an extrapolation of the ethnicity statistics based on the 1930 Colonial Census. Another source of information is the statistics on languages spoken at home, as collected during the 1990 Population Census. According to the 1990 Census, the composition of languages spoken at home was as follows: Javanese (40%), Sundanese (25%), Indonesian (13%), Madurese (4.3%), Minangkabau (2.3%), Batak (2.2%), Bugis (2.2%) with the rest, minority languages. The estimation of ethnic composition using language as a proxy is very unreliable due to the fact that a relatively large proportion uses the national language rather than their local language; also, as people migrated they could possibly have replaced their original language with their new local language. In 2000, surprisingly, the government decided to include ethnicity in the Census questionnaire.

important part of the state policy and strongly characterizes the political demography of Indonesia.[3]

Indonesia, however, is not unique in its attempt to eradicate ethnic plurality, as Esman (1994) clearly argues: ... *as the duty of state elites was to build a united and homogenous nation, the duty of ethnic minorities was to assimilate; the only acceptable alternative to assimilation was passivity.* According to Esman (1994), ethnic pluralism can be traced to three factors: First is conquest and annexation when a people is defeated and brought under the rule of the victor; second is the process of European colonization and decolonization which assembled within administrative boundaries, established for the convenience of the colonial powers, people who had no mutual affinity and were often split among two or more states, governed by different colonial masters; and third is a population movement, as people cross political boundaries, in search of economic opportunities, or religious and political freedoms. For Indonesia, the source of ethnic pluralism, using the Esman categories, is: Firstly, the process of colonization and decolonization, which (re)-constructed the existing ethnic boundaries; and secondly, the population mobility, both internally and internationally, which provided a public arena for ethnic integration and disintegration. The discourse on ethnicity or, more broadly, on cultural pluralism, in postcolonial Indonesia, can be attributed, therefore, to the effort to unravel the colonial legacy and, at the same time, to construct a new national identity, based on the Indonesian founding fathers' idealized and imagined community. Ironically, the postcolonial state, as exalted by the Indonesian founding fathers, represents the state that was configured by the colonial state. Lindbald (2002, 93–94) has succinctly elaborated the development of the colonial state—the predecessor of the current Indonesian nation-state, as follows:

> *The emerging of Dutch colonial state, its authority unchallenged in Java after the conclusion of the Java War (1825–1830), was gradually extending its control over the rest of the Indonesian archipelago. It was a long process, facilitated by the political fragmentation of the Outer Islands. To all intents, the only modern state in the Indonesian archipelago during the nineteenth century was the Dutch colonial state with its center in Java.*

After Independence, paradoxically, the national elites perceived colonialism as having used a "divide and conquer" strategy to weaken the growing aspirations among the Indonesian population to unite and strive for national independence.[4]

[3]Demographic engineering, a term I borrowed from Bookman (1997), is broadly defined as the state attempt to reconfigure the ethnodemographic composition of the population for various strategic purposes. Bookman, M. Z., 1997, *The Demographic Struggle for Power: The Political Economy of Demographic Engineering in the Modern World.*

[4]In an interview, Bernard Dahm, a historian who wrote the political biography of Sukarno, the first Indonesian President, recalled his meeting with Sukarno: *I remember vividly, I had a long talk with Sukarno in October 1966. I have always seen his attempt to unite the people in the old tradition of the Javanese, a syncretic attempt to synthesize great influences from outside. Sukarno was romantic and had seen that nationalism, religion and communism (Nasakom) worked well in the 1920s and 1930s. He believed he could do that in the 1960s and I said that is an illusion* (*Jakarta Post*, 29 March 2001).

Acknowledgments

The ideas and a large part of this book have traveled a long way from their inception more than two decades ago in the tranquil village of Wassenaar, where NIAS is located. At that time, I was lucky to be part of the 2000–2001 group of NIAS fellow-in-residence in which the first draft of the book germinated. It took almost a decade before the draft was published by LIPI Press in 2013. This book is a revised version of that book. Between 2013 and 2021, Indonesia has experienced many changes that some parts of the first publication need to be adjusted. The major change is the relocation of one chapter (Chap. 11 becomes Chap. 6) and the replacement of two chapters (Chaps. 5 and 8). The reason for relocation and replacement with the new article is mainly for making the book more coherent and concise. The book aims to provide a long history of the interface between demography and politics in a country that represents the largest archipelago and the fourth largest demography in the world. In this revised version, an epilog is added entitled "Indonesia in the Twenty-First Century," a brief reflection on the future.

On the long journey that finally has ended up in this current publication, I incurred a lot of debt to many individuals and institutions. I would like to especially thank Thee Kian Wie as due to his recommendation, I could stay as a 2000–2001 fellow-in-residence at NIAS (the Netherlands Institute for Advanced Study in the Humanities and Social Sciences). Henk Wesseling and Jos Hooghuis, the rector and the executive secretary of NIAS, generously provided facilities that made my life and work pleasurable and which I greatly appreciated. As some of the chapters were originally written during my doctorate studies at the Australian National University (ANU) in 1986–1990, I would like to thank my supervisors and advisors, Gavin Jones, Paul Meyer, Lorraine Corner, Thomas Smith, and Geoffrey McNicoll, for their remarkable guidance.

My sabbatical as Fulbright Fellow at Brown University in 1996–1997 exposed me to the wider perspective of migration, ethnicity, and nation-building and the admiration of the collegiality of Sidney Goldstein, Calvin Goldscheider, Dennis Hogan, and David Ketzer. In 2003–2004 I had the opportunity to stay as a visiting professor at the Institute for Language and Culture of Asia and Africa (ILCAA) at

Tokyo University for Foreign Studies, where I could continue working on this book. I gratefully acknowledge Koji Miyazaki, the director of ILCAA, for his friendship and support. From 2005 to 2007, my research collaboration with Stephen Castles at the Refugee Studies Center and Francis Stewart at the Center for Research of Inequality, Human Security and Ethnicity both at Oxford University further advanced my horizons and delighted me greatly. Most parts of this book were finally completed during my 3 months' stay as a senior research fellow at ARI (Asia Research Institute) in Singapore in 2008, where support from Gavin Jones, my former guru at ANU, and Brenda Yeoh, the leader of the migration cluster at ARI, was deeply appreciated.

As the last chapter in the first book was originally written in 2005 while the main ideas in the book still relevant today, a necessary adjustment needs to be made to make sense for the current readers. Although the ideas and most parts of the book have traveled to many places in the world it would never have materialized without the guardianship of individuals and institutions back in Jakarta. I, therefore, dedicate the book to LIPI (Indonesian Institute of Sciences), my research home base that has nurtured me for almost four decades. I gratefully acknowledge Leknas LIPI (1980–1984), where I began my career as junior researcher, Research Center for Population Studies (1985–1998), and Research Center for Social and Cultural Studies (1999–2017) as the book would never have been drafted without the constant support of my research institution.

My research colleagues' names could not be mentioned one by one here, but together we make a warm research community that has greatly contributed to my career as a LIPI researcher. In the final stage of preparing the revision of the book, I am incredibly grateful for the assistance of and full support of LIPI Press, Springer, and two anonymous reviewers who have given endorsement for the revised publication of the book. This book is hopefully not only meaningful for me as the author but, more importantly, for the society at large. The revision that is conducted in the middle of the Covid-19 pandemic reminds all of us that solidarity and interconnectedness should be even more enhanced to overcome the universal challenge of human suffering.

Jakarta, June 2021.

Contents

About the Author

Riwanto Tirtosudarmo accomplished his doctoral degree in social demography at Research School of Social Sciences, the Australian National University, Canberra, Australia, in 1990. Since 1980 he works at the Indonesian Institute of Sciences (LIPI) as a social researcher and retired in 2017. He publishes extensively in scientific and popular journals on political demography and the politics of population mobility in Indonesia and Southeast Asia, and is the author of *Demografi-Politik Pembangunan di Indonesia: Dari Riau sampai Timor Timur* (The Political Demography of Development in Indonesia: From Riau to East Timor) in 1996; *Mencari Indonesia: Demografi Politik Pasca-Soeharto* (Searching for Indonesia: The Post-Soeharto Political Demography) in 2007; *Mencari Indonesia 2: Batas-Batas Rekayasa Sosial* (Searching for Indonesia 2: The Limits of Social Engineering) in 2010; *On the Politics of Migration in Indonesia and Beyond,* 2016; republished as *The Politics of Migration in Indonesia and Beyond* in 2019 (Springer). He was a Fulbright Visiting Fellow at the Department of Sociology, Brown University (1996–1997), Fellow-in-Residence 2000–2001 at the Netherlands Institute for Advanced Study in Humanities and Social Sciences (NIAS), Visiting Senior Research Fellow at Refugee Studies Center, Oxford University (2002), and Visiting Professor at the Research Institute of Language and Culture, Tokyo University of Foreign Studies (2003–2004). In March–April 2008, he was visiting Researcher at KITLV, Leiden, and Visiting Senior Research Fellow at Asia Research Institute,

National University of Singapore (July–September 2008). He served as a member of the International Editorial Advisory Board of *Journal of Immigrant & Refugee Studies* and *Asian Policy and Politics*.

Part I
Demographic Engineering and Territorial Integrity

Geography had a strong influence upon the course of Indonesia's history and has always conditioned severely the course of political development.
George McT. Kahin (1965, 535)

The Indonesian government has explicitly stated that transmigration is an instrument of population distribution policy. Transmigration is defined as a government policy that involves the clearing and opening of land that was previously uncultivated or only sporadically or illegally cultivated, and the deliberate movement and resettlement of people. The idea of population transfer from Java to other sparsely populated islands is far from new. This policy has been in force since the beginning of the twentieth century.[1] The island of Java is renowned as one of the most densely populated areas in the world. The problem of population, particularly its high growth and its large size, has become more pressing as it has coincided with the problem of poverty among a large part of the population. From the beginning, transmigration has been perceived as a policy to solve simultaneously the demographic problem and the welfare problems of the population of Java. From a demographic perspective, transmigration policy has had little impact on the reduction of the population in Java. Assuming, for example, that the government target to resettle 150,000 families a year during the Fourth Five-Year Development Plan (1984–1989) had been realized, and assuming that every family consisted of five persons (two parents and three children), only 750,000 persons a year would have been moved.[2] This number, taking the 1980 annual growth rate of 2%, and the size of the population of Java and Bali as approximately 100 million, suggests that a successful transmigration policy could have removed only just over one-third of Java's annual population growth and would not have prevented Java's total population from continuing to increase. It also is evident that the unbalanced population

[1] Transmigration, far from being an exception, is the archetype of resettlement policies. For an interesting comparison with other population resettlement policies in the world, see Levang (1995).

[2] This is the highest target of the transmigration policy to date.

distribution between Java and the other islands would have been altered only marginally by transmigration.

While in the long term the transmigration program has had only a marginal effect on population growth in Java and little influence on population distribution in the archipelago, it has had a tremendous demographic impact on the receiving areas. Lampung and South Sumatra are classic examples of the remarkable contribution of transmigration in accelerating regional population growth. In other provinces, although the overall demographic impact of transmigration has been less remarkable by comparison, it has been significant for some districts. In 1985, for example, 25% of the population in the district of Kampar in the Province of Riau was transmigrants. In the district of Kotabaru in the Province of South Kalimantan, more than 40% of the population was transmigrants.[3] After Independence in 1945, the relationship between the central and regional governments was a crucial aspect of national integration and development.[4] Since the beginning of New Order in 1966, the Indonesian state can be described as strongly centralized in character, for example, nearly 80% of national expenditure is planned and disbursed by the central government.[5] The transmigration program is perhaps the government policy that most reflects the strength of the central government in determining affairs in the regions, as it has been almost totally planned by the central government in Jakarta. The regional governments, whose lands were designated as receiving areas for settlers from Java and Bali, were basically passive participants in the planning and implementation processes.[6]

After the mid-1980s, the transmigration policy basically lost its economic rationale as economic opportunities grew more rapidly in Java than on the outer islands. The macroeconomic shift instigated by the collapse of oil prices forced the state to adopt a more export-oriented policy, replacing the previous import substitution strategy. The transmigration program, within the macroeconomic context, therefore, became a counterproductive policy, as the overall development trend clearly promoted the creation of employment opportunities in Java rather than in the outer islands. Surprisingly, however, the transmigration policy is still highly regarded as a national priority on the state's policy agenda. The perception among the ruling elite, particularly the military, that the transmigration policy is pivotal to national security is arguably the strongest factor in supporting transmigration's continuation on the state policy agenda. In this part, there are three chapters that analyze the transmigration policy as the state's demographic engineering instrument for strategic and security purposes. Chapter 1 describes the historical continuity of migration policy, from the Dutch's emigration policy to the Indonesia's transmigration policy.

---

[3] World Bank (1988, 21).

[4] See studies on the center-regional relations, among others by Maryanov (1958), Legge (1961), Nawawi (1968), Bachtiar (1974), Mackie (1980), Achmad (1984), MacAndrews (1986).

[5] The New Order regime finished with the stepping down of President Suharto on 21 May 1998 in the midst of the severe economic and political crises.

[6] See Tirtosudarmo's study on "transmigration and its center-regional context" (1990).

Chapter 2 specifically analyzes the place of transmigration policy within the context of development planning in the New Order's period. While Chap. 3 will be looking at how transmigration policy is implemented in two provinces, namely Riau and South Kalimantan, to expose the center and regional context of this policy. To demonstrate both the continuity of and changes in the policy, the chapter adopts a chronological approach, tracing the history of the policy since its early formation at the turn of this century until the end of Suharto's New Order period in May 1998.

Chapter 1
From *Emigratie* to *Transmigrasi*

The Making of an Emigration Policy

The end of the nineteenth century and the beginning of the twentieth century were marked by several important social changes, the impact of which significantly altered the course of history in Indonesia. Although each can be considered as a single event, they can be more usefully regarded as the complexes of interconnected events. Ten years after the Dutch had recovered from the Diponegoro War (1825–1830), they began to pay attention to territories outside of Java (the Outer Islands). From about 1840 onwards, Dutch involvement expanded throughout the Outer Islands. According to Ricklefs (1981, 125), there were many reasons for this expansion; however, two general considerations applied everywhere: First, to protect the security of areas they had already held, the Dutch felt compelled to subdue other regions that might support or inspire resistance movements, and second, as the European scramble for colonies reached its height in the late nineteenth century, the Dutch felt obliged to establish their claims to the Outer Islands of the archipelago to prevent other Western power from intervening, even the Dutch themselves initially had no interest.

Ricklefs (1981, 138) noted that "*By about 1910, the boundaries of the present state of Indonesia had been roughly drawn by colonial armed forces, at a great cost of lives, money, devastation, social cohesion and human dignity and freedom.*" At the dawn of the twentieth century, Dutch colonial policy showed a significant change in direction. Its jurisdiction was expanding as the invasion of the entire archipelago neared completion. It also took on a new mission: The direct exploitation of Indonesia that peaked during the culture system (*cultuurstelsel*) in 1830–1870 gradually became less important to the Dutch who began to accept responsibility as colonizers for the welfare of Indonesians. One important element of the new policy was the introduction of a native welfare program, the nature of which is illustrated by the slogan of the prominent colonial reformer van Deventer:

R. Tirtosudarmo, *From Colonization to Nation-State*,
https://doi.org/10.1007/978-981-16-6437-3_1

"irrigation, emigration and education" (Penders 1969, 27). This policy has become well known as the "ethical policy."

The genesis of the ethical policy, as elucidated by Ricklefs (1981, 143), was the simultaneous outcome of the social, political, and economic transformation experienced by Dutch society in the Netherlands:

> *The Ethical policy had roots both in humanitarian concern and economic advantage. The denunciation of Dutch rule presented in the novel Max Havelaar (1860) and in other exposes began to bear fruit. Voices were raised in favour of relief for the oppressed peoples of Java, and by the late nineteenth century new colonial administrators were on their way to Indonesia with Max Havelaar in their baggage and its message in their minds. During the "liberal" period (c.1870–1900) private capitalism came to exert a preponderant influence upon colonial policy. Dutch industry began to see Indonesia as a potential market, which required a rising of living standards there. Capital, both Dutch and international, sought new opportunities for investment and extraction of raw materials, especially in the outer islands. A need for Indonesian labor in modern enterprises was felt. Business interests therefore supported more intensive colonial involvement in the causes of peace, justice, modernity and welfare. The humanitarians justified what the businessmen expected to be profitable, and the Ethical Policy was born.*

Java's rapid population growth after the middle of the nineteenth century was one of the dominant facts of the colonial age. The central and eastern parts of Java were becoming seriously overpopulated, while in the Outer Islands there were still vast areas of unpopulated or underpopulated land. Java's growing population was considered by the Dutch government to be the underlying cause of the decline of its people's welfare. The Dutch, however, had no policy that could directly solve the problem of population growth. The only answer they offered was emigration from Java to the Outer Islands. As Ricklefs (1981, 154) correctly argued,

> *There was no real attack on population growth. The nation's wealth was used for the interests of foreign enterprises and indigenous industries were not developed. The main economic development took place in the outer islands, while the main welfare problems grew in Java. Education was to be the key to a new age, but the number of school places provided was small when set against the size of the population.*

In a similar vein, van der Kroef (1956, 743) noted that the schemes devised by the colonial government to cope with the dangerous economic effects of population pressure, curiously enough, rarely concerned themselves with the structural transformation of agricultural production. Many officials of the colonial and the postcolonial government as well as some scholars have based their hopes on emigration as a solution of the problem. Pelzer (1945a, b), a noted geographer, for example, has even gone so far as to maintain that emigration from Java to less densely populated islands will be able to absorb the population increase for many decades and that emigration rather than industrialization would provide the necessary relief.

From a different perspective, Kartodirdjo (1973) perceived that the rural history of Java in the nineteenth and early twentieth centuries was marked by sporadic movements of peasant unrest. Many of these peasant unrests climaxed in violent clash with the colonial authorities. Although a direct link between social unrest in rural Java and the initiation of the emigration policy cannot be established, it is not implausible to posit a cause–effect relationship. A major complicating feature of agrarian unrest in Java, according to Kartodirdjo (1973, 1), had been its correlation with social change in general and with the colonial impact in particular. He explained that.

> *The social movements had as their background, the rapid penetration of a colonial economy whose impact on rural Java reached a climax during the course of the nineteenth century. The colonial rulers introduced a new legal and social relationship covering agrarian and labor matters. Excessive demands for services and the levying of new taxes exacerbated popular discontent. The turning over of some land to sugar cultivation and the exaction of a compulsory contribution of paddy had a direct bearing on some of the instances of social unrest.*

According to Penders (1969, 28), the ethical policy of indigenous economic development was almost entirely concerned with raising agricultural productivity and the Dutch apparently had no serious intention of introducing drastic changes in the colonial economic structure by such means as large-scale industrialization. As Legge also argued, the measures taken by "ethical" colonial governments to improve the living standards of the indigenous population were "*In the nature of palliatives. They provided some alleviation of particular areas of hardship, but they did not achieve—and they did not attempt to achieve—any thoroughgoing technological changes...*" (Legge 1964, 91 cited in Penders 1969, 28). Penders (1969, 28) also noted that the importance of large-scale industrialization was stressed by a number of ethical reformers, but until the worldwide economic depression of the 1930s the various plans that were suggested were not implemented. The Calvinist Prime Minister Kuyper (cited in Penders 1969, 29) even stressed in 1901 that the only way to develop the Indies was "*to lift the people of the Indies from the agricultural to the industrial state.*"

According to Penders (1969, 28), the Minister of Colonies, Idenburg, in fact strongly supported Kuyper and envisaged the establishment of indigenous industries owned by indigenous capital. On another occasion, the socialist van Kol dismissed van Deventer's view and stated that.

> *The expectations of van Deventer that emigration is the proper means to create a balance between production and consumption are delusive and will never come true. I believe that his plan for salvation which is indicated by the trilogy: "education, irrigation and emigration" is inadequate because the distress has taken on too vast proportions. He believes that in particular indigenous education will have an enormous impact on the well-being of the Javanese. I am of the opinion that this is an illusion which is completely out of touch with reality. The only solution is a more intensified production and this is only possible with industry, preferably large-scale indigenous industry...* (cited in Penders 1969, 29).

Van Oorschost (cited in Penders 1969, 29) mentioned the following reasons for the Dutch lack of interest in industrialization: First, the interior freight rates were too high; second, a widespread distribution apparatus was missing; and third, import duties on industrial goods were too low to afford sufficient protection. Penders (1969, 29–30) argues, however, that Van Oorschost does not mention—or at least does not spell out fully—the fundamental reasons for the failure of the attempts at industrialization:

> *One most important cause of this failure was the strong opposition of the politically powerful plantation combines. Large-scale industrialization would have tended to increase the demand for labor and therefore the price of labor. This would have been damaging to the plantation concerns which—as was pointed out above—were largely dependent for their profits on low wages. But also Dutch industrialists as well as Dutch labor were unwilling to be priced out of the Indonesian market which was one of the most important outlets for Dutch industrial product. In the final analysis then the industrialization of Indonesia militated against the interests of the imperial economy as a whole.*

In serving the interests of plantation companies to control their labor, in 1880 the Netherlands Indies' government issued the first Coolie Ordinance to regulate labor relations, particularly in East Sumatra. This ordinance was later expanded to other regions in the Outer Islands. It protected the interests of plantation companies in retaining the workers for the duration of the contract period. A "penal sanction" was imposed on workers who violated their labor contracts. The first Coolie Ordinance was followed by other ordinances in 1884 and 1893, which provided employers with effective legal control over their indentured workers. In addition to the penal sanction, employers resorted to other means to keep their workers. One favorite way was to provide gambling opportunities for the workers on pay day. Often workers become so deeply indebted that they had no other choice but to sign a new contract with their employers (Thee 1969, 7–12). According to Thompson (1947, 134), until 1909 the recruitment of Javanese estate labor likewise remained in the hands of professional agents, but in that year, planters were encouraged by law to appoint their own agents who were often aided by worker recruiters. The Deli Planters' Association, for example, led the way by setting up its own agency in Semarang, and direct recruiting by employers gradually became widespread. In 1930, it was found that 84% of the labor employed in plantations on the Outer Islands was Javanese. Of the remaining, 13% was Chinese and 3% was locals.

Emigration as an attempt by the colonial government to move people from Java to the other islands is considered the least successful measure of the ethical policy. The first organized attempt in this direction was made in 1905 when, as an experiment, an agricultural colony called Gedong Tataan was set up in Lampung (South Sumatra). By 1930, the colonists numbered 30,000. Efforts to settle Javanese farmers in the southeast of Kalimantan and Sulawesi met with failure. The situation improved somewhat during the 1930s when, as a result of more skillful propaganda, better selection methods, and more extensive preparatory work in the areas of settlement, more farmers could be induced to leave.

Following previous writers such as Pelzer (1945a, b), Heeren (1979, 10–15) divided Dutch resettlement policy (1905–1941) into three broad phases. First, there was the experimental phase (1905–1911). The assumption adopted during this phase was that new settlements should, as far as possible, resemble villages in Java. Up to 1911, only about 4,818 Javanese had been moved to Lampung. Second, came the period of the Credit Bank of Lampung (1911–1928). The government established a Credit Bank of Lampung in March 1911 to provide 22.5 gulden as a migration premium to every migrant family, in addition to a maximum of 300 gulden, given as credit. This amount had to be repaid by the migrants after 3 years with 9%

per annum interest. Migrants could borrow additional money for the purchase of, for example, livestock, agricultural equipment, or building materials. Since the majority of the migrants did not have any experience of banks, problems of repayment were inevitable. Many used the money for unproductive purposes, such as for clothing or festivities *(slametan),* or even for alcohol or drugs *(candu).* Another problem during this phase was internal mismanagement and corruption within the bank. In 1926, the bank was heavily in debt, and by 1928, the government had decided to close.

In the second phase, the problems were also caused by the selection system, which was based on a target system: The *lurah* (village head) was required to meet a designated target number of migrants each year. Consequently, many migrants did not meet formal selection criteria; some were selected by the *lurah* merely because he wanted to remove them (e.g., the old, the sick, or landless laborers) from the village. The emigration policy during this period was also challenged by the plantation interests, particularly from the agency responsible for recruiting contract labor from Java. They spread rumors to discredit the settlement in Gedong Tataan, in order to attract migrants to work on their plantations, mostly in Deli (North Sumatra). Despite these many problems, a second settlement area, Wonosobo, was opened in Lampung in 1921 and by 1924 this village had a population of 5927. The population in Gedong Tataan had risen to 25,000 by 1928. According to Heeren (1979, 12), by the end of the twenties the numerous problems that had been experienced brought the whole idea of the emigration policy into question. Conditions in the settlements were considered by many observers as far from satisfactory. In addition, the impact of the 1930s economic depression reduced the availability of financial resources for the program. The thousands of returned migrants who were formerly employed in the plantations were seen to aggravate the already alarming population problem in Java.[1] Thus, migration on a large scale to the Outer Islands was favored by the government (Perez-Sainz 1983, 194).

In the mid-1930s, the plantation companies, most of which were in East Sumatra, were experiencing many difficulties in recruiting labor. The difficulty in recruiting labor is mainly because the government due to the political pressure had to abolish the penal sanction. With the abolition of the penal sanction, many dissatisfied Javanese workers had returned to Java. Thus, the gradual success of the agricultural colonies of Javanese settlers in the Outer Islands was seen by the planters to constitute an increasingly important source of labor (Thee 1969, 78). The government's need to reduce the population growth in Java was compatible with the interests of the planters in recruiting cheap labor. Thompson (1947, 135) has argued that the official agricultural colonization program in the Outer Islands was designed primarily to help solve the labor shortage. The government commenced the third phase of the emigration policy, known as the *Bawon* system. The *Bawon* system was initially devised to meet the needs of migrants in Lampung for labor to harvest their rice. Under the *Bawon* system, the government only provided land for the migrants.

[1]Thousands of migrants returned to Java as many plantations had to reduce their production after the world economic crisis in 1930.

They were to pay back their own fares within 2 or 3 years of settlement. The government usually moved them into Lampung from Java during harvest time, around February–March, so that they could be hired immediately by earlier migrants. Their wages were paid in kind, in the form of rice. In Java, their share of the harvest typically was one-tenth; in Lampung, they could earn as much as one-seventh or one-fifth of the harvest. From this, they paid their debts and started a new life as colonists in Lampung. The *Bawon* system was considered the most successful emigration program of the Dutch period, moving large numbers at relatively low cost. In addition to those settled under the *Bawon* system, between 1935 and 1940 the government also sent people to new settlements in other areas of Sumatra outside Lampung and to Kalimantan.[2] As also noted by Nitisastro (1970, 90), after 1931 the government migration policy experienced a number of changes. In 1937, a special commission in charge of the migration policy was established. According to Nitisastro (1970, 90), after those policy improvements, the annual number of migrants increased to 14,700 in 1935 and 52,800 in 1940.

The Dutch's government migration policy (*emigratie*), as one of the trilogies of ethical policy objectives, was essentially formulated and maintained as the result of a combination of forces. First was the political change in the Netherlands, which brought the Calvinist–Catholic coalition to power in 1901. The outstanding feature of the policy outcomes from this new coalition was the official abandonment of the goal of economic exploitation and the introduction of direct intervention in the economic sphere to improve the position of the indigenous population. The second factor was economic opportunity, particularly as seen by the Dutch capitalists after the Dutch had successfully conquered the whole archipelago. The vast land areas in the Outer Islands attracted private companies to open plantations. Given the scarcity of labor, Javanese were recruited as cheap labor for new economic activities. The third factor was social and political unrest in many parts of rural Java, resulting from simultaneous economic exploitation and population pressures. These in turn encouraged the colonial policy to attempt to deal with social unrest through population movement to the Outer Islands. Emigration policy, as developed by the Dutch, was a useful instrument for serving many goals and interests.

[2]From 1933 to 1941, 222,586 migrants were resettled in Lampung and South Sumatra (Heeren 1979, 14), and almost 2500 migrants in Kalimantan (Pelzer 1945a, b, 223–225). During the Japanese occupation, the government's migration policy was practically halted. The economic situation in Indonesia also deteriorated. The population continued to increase but the rate of increase in economic productivity fell seriously behind that of population, owing to the destruction of the economic apparatus during the Japanese occupation and the revolution.

Transmigration: Migration Policies After Independence, 1945–1965

Among the new national leaders, Hatta (1954, 169–170) argued most forcefully that emigration should be continued after independence. However, the idea of "transmigration" as proposed by Hatta was somewhat different from the colonization policy implemented by the Dutch. Transmigration was to be implemented in conjunction with the industrialization of the Outer Islands. Hatta's ideas on transmigration were understandable. The superiority of industrialization over transmigration was an old idea among socialists in the Netherlands who had strongly influenced his thinking. Hatta's political and economic ideas were crystallized during his period as a student in the Netherlands in the 1920s when socialism was strong in Europe. The socialist politicians in the Netherlands demanded a genuine process of industrialization to improve the economic conditions of the indigenous people, rather than merely the emigration of Javanese to the Outer Islands, "*Without industrialization, transmigration will never solve the problems. Transmigration will only postpone the problems that eventually will return sometime in the future. Transmigration settlements, therefore, should be formed simultaneously with industrial locations.*"[3]

Hatta strongly criticized the colonial policy, which had, basically, only shifted landless farmers from Java to become farmers in the Outer Islands. He argued that,

> *Transmigration, which had been conducted by the Dutch, only resulted in shifting peasants to be peasants. In new settlements, those peasants established their villages based on traditions in Java. Even their house construction is also a replication of their customs. Everything is done primitively. Their spirit is static. In their former villages their conditions were very bad; in the new villages, they continue their backward way of living. This way of life, which does not bring any progress, will continue in the new settlement. There is no fresh spirit. Their lives have only postponed the coming problems, until the time when their families are grown-up. Therefore, wherever they go they will be caught up by the misery which they left behind.*[4]

Hatta (1954, 171) also suggested that "*'Emigratie' or 'transmigrasi' will achieve its goal in opening up prosperity if the people are given guidance and technical assistance in their new environment. If they are ignored and neglected, their life will never change. So far, all efforts have been made with a very narrow mind. Therefore, sooner or later, they will be people 'who eat land'. Their family members will grow up and the village, therefore, will need to be extended, and it will not be surprising that they will encroach on every available piece of land*" (author's translation). After independence, the so-called *Panitia Siasat Ekonomi* (Committee for Economic Strategies), of which Hatta was one of the masterminds, formulated a plan in which industrialization and population resettlement were the two major

---

[3]Hatta (1954, 169), author's translation.

[4]Hatta (1954, 170), author's translation.

components.[5] Thompson (1947, 179) who observed the first Indonesian Government plan noted that "*The resettlement program, too, is an amplification of a prewar project—with certain essential differences. Sumatra, and eventually others of the under populated Outer Islands, is to receive immigrants from Java. Individuals, not entire villages, will be selected for their youthful vigor and initiative and will be transported in conjunction with the industrialization schemes for the Outer Islands.*"

However, the idea of industrialization as the backbone of Hatta's transmigration policy proved unattainable. As Wertheim observed in 1956, the difference between the Dutch and the Indonesian resettlement policy had nothing to do with industrialization,

> *Before the war Dutch resettlement policy was directed towards the creation of typical Javanese enclaves in a purely Sumatran environment. As the willingness to move to 'the land beyond' had not yet developed very greatly at that time, the Dutch tried to make the settlements attractive, by reshaping conditions as nearly as possible to resemble those in the homeland. After independence, the Indonesian Government adopted a different view. The strict separation of Javanese and Lampongers was deemed out of date, as a relic of colonial policy. Such a system would not fit into the new ideology of Indonesia as a unitary state. Javanese and Sumatrans were equally Indonesians, who had to live together and mix on an equal footing. Both population groups would have to develop into modern Indonesians and work together in building a new society in the melting pot of Sumatra, the Land of Promise* (Wertheim 1959, 149).

It is striking that the office in charge of implementing resettlement policy after independence had itself constantly "migrated" from one ministry to another. The unsettled political climate following independence, as the Dutch attempted to regain their former colony, and an unpreparedness of the Indonesian leaders to lead the country were contributing factors in the failure to establish a single solid implementing agency for transmigration.[6] The first transmigration office came under the Ministry for Labor and Social Affairs. In 1948, it was shifted to the Ministry for Development and Youth Affairs. Later in that year, the ministry was abolished, and transmigration was transferred to the Ministry of Home Affairs. In January 1950, the government established a separate office for transmigration (*Kantor Transmigrasi*), which in October was placed under the Ministry of Social Affairs. It was under the auspices of this ministry that the first group of transmigrants was resettled in Lampung.

In 1951, the Head of the Transmigration Office, Ir. A.H.O. Tamboenan, formulated a plan to resettle 10,000 families to South Sumatra and Sulawesi each year. Because of financial stringency, only 653 families were moved, and most of these were resettled in areas adjacent to the former Dutch colonization sites (Sjamsu 1960,

[5]The fact that industrialization and transmigration became an important part of the government plan immediately after independence was, according to some observers, rather strange. Newly independent Indonesia had been born of a rhetoric that called for the rapid destruction of colonial and economic institutions (Paauw 1963, 155).

[6]Although the government established an office to implement transmigration policy in 1947, the first group of transmigrants (about 23 families) was only resettled in 1950 in Lampung (Hardjosudarmo 1965, 127).

80). Approximately 26 families were resettled in Sulawesi also close to a former area of colonization. These transmigrants had come from the regencies of Purworejo, Kebumen, Banyumas, and Probolinggo, Central Java. The transmigration office also resettled 114 families of refugees from East Priangan to Banten, West Java, during this time. They were evacuated because their villages had often been terrorized and ransacked by an Islamic rebel group (Darul Islam) operating in the region. Under the Ministry of Social Affairs, the stated goal of transmigration was to improve the prosperity and welfare of the population, through the development of the economy in various sectors (Hardjosudarmo 1965, 128). In 1952, Tamboenan proposed an ambitious 15-year plan for transmigration. According to the plan, within 1 year the government would open up 30,000 hectares of land in Sukadana in South Sumatra. In conjunction with the clearing of this land, the government planned to send 1000 families to be employed there. Moreover, the government also established a land clearing agency called Japeta (*Jajasan Pembukaan Tanah*). In the end, however, this plan achieved little. Only 3855 hectares of land was cleared, and 3.851 families were moved, mostly to South Sumatra. The so-called Tamboenan plan slowly disappeared.

In the fifties, in addition to the so-called general transmigrants (*transmigrasi umum*) there were four other types of government-sponsored transmigrants. They were "S. O. B." prisoners, Surinam repatriates, ex-guerilla fighters, and ex-army members. The "S. O. B." prisoners were suspected by the government of being involved in the Islamic rebel group in West Java during the early fifties.[7] Based on Presidential Instruction No. 54, 1954 (Sjamsu 1960, 64), such people and their families were sent to transmigration settlements in Sumatra and Kalimantan. In 1954, approximately 500 families were resettled in Pangkalan Susu in North Sumatra, and another 242 families in Samboja in East Kalimantan. In 1955, a further 326 families were sent to those two regions (Sjamsu 1960, 94).[8] When the war with the Dutch came to an end, many ex-guerilla fighters who had lived in the various stronghold areas to fight the Dutch were confronted by serious problems of adjustment when they returned to their former homes. In February 1951, the government established a National Reconstruction Office (BRN) under the Ministry of Home Affairs to assist them: *The government feels that job opportunities have to be created for various groups of ex-guerilla fighters. The aims are, first, to assist them in the adjustment to the normal society; and second, to utilize their potential to rebuild the nation, in ways that are economically justifiable.*[9]

[7]"S. O. B." (*staat van oorlog en beleg—Dutch*) means "dangerous areas" (*daerah bahaya*—Indonesian).

[8]In 1954, the transmigration office resettled almost 300 families, mainly of Javanese origin, who had just returned from Surinam, another Dutch colony in the Caribbean Islands. They were Indonesians who had been recruited by the Dutch to work as laborers in Dutch sugar plantations in Surinam. The government resettled these people in Tongar in the regency of Pasaman Barat, West Sumatra.

[9]Government Decree No. 12/1951, cited in Heeren 1979, 68, author's translation.

From the early 1950s, agricultural activities were considered the best solution to "re-socialize" ex-army personnel and ex-guerilla fighters. Given that further land for agricultural extension was not available in most parts of Java, resettlement in the Outer Islands was seen as an appropriate strategy. Another benefit from this scheme was the removal of ex-guerilla fighters from "an unhealthy environment" in the capital city. Around 10% of the ex-guerilla fighters who became transmigrants had been born in Jakarta (Heeren 1979, 71). Another reason for this scheme was to protect such people from being attracted to Islamic rebel groups in West Java.[10] Most of the projects under this scheme were implemented during 1951 and 1952. By 1957, there were almost 22,000 migrants in the new settlement areas, mostly in South Sumatra. During this period, the government also introduced and promoted various kinds of agricultural experimentation through cooperative and collective systems. The resettlement scheme for ex-soldiers was similar to that for ex-guerilla fighters. Ex-soldiers also experienced adjustment problems when they returned to civilian life. Under the auspices of the Ministry of Defense, a government agency called Bureau for the Resettlement of Ex-military Members (BPBAT) was established to open land for plantations and agriculture and provide better economic opportunities for ex-army transmigrants. From 1950 to 1955, the government successfully resettled more than 6000 ex-soldiers, mostly in Lampung.[11]

In the period 1956–1960, the government formulated for the first time a five-year development plan, in which transmigration was described as an instrument to reduce population pressure in Java; provide labor in the sparsely populated provinces; and support military strategy, as well as to accelerate the process of assimilation (Hardjosudarmo 1965, 128–129). The most significant transmigration policy chance in the plan was the explicit reference to a strategic military goal. This was apparently due to increased political unrest in some regions, particularly because of disappointment with central government leadership. The important strategic role of transmigration was further emphasized in 1962, as a result of the change in the national constitution (Hardjosudarmo 1965, 130). President Sukarno declared the new constitution to be a "guided democracy," which replaced the "liberal or parliamentary democracy" that he considered a failure. The establishment of an autocratic "guided democracy" and "guided economy" under Sukarno, according to Kuntjorojakti (1978, 138–139), was always with the support of the central military leadership. The political role of the military increased considerably after they had successfully destroyed the regional rebellions between 1956 and 1958 in West Sumatra, North Sulawesi, and South Sulawesi. As Feith and Lev (1963, 37) argue,

[10]This fear was realistic. More than 70% of the ex-guerilla fighters who resettled in South Sumatra originally came from West Java, mostly from places around Jakarta, like Bogor, Bekasi, and Karawang (see Heeren 1979, 71–72).

[11]According to Sjamsu (1960, 88), it is difficult to measure the statistical achievement of the transmigration policy in South Sumatra. The transmigrants mixed with previous settlers, and many migrants who moved to South Sumatra either came voluntarily or under various other government schemes. One thing was certain, however, as many observers had argued: Lampung was becoming another Java.

After the end of the PRRI rebellion, through the removal of the regionalists from the political and military scene, the rebellion cleared the way for the emergence of a new set of power relationships. Of the three main political forces of 1957 (the President, the central army leadership, and the regionalists) only the first two remained by mid-1958, and these continued to be markedly distinct and, in many ways, competitive with one another. But they also had many goals and interests in common.

The course of the regional rebellions had caused the government to adopt more centralized attitudes in policy. According to Kuntjorojakti (1978, 139),

At the national level, several Presidential edicts on regional matters were issued around 1959–1960. These decrees treated the regions as mere administrative units within the framework of a highly centralized national bureaucracy, totally subservient to the will of the central regime. This fitted very well with Sukarno's idea of a monolithic Unitary State, which was redefined as a political instrument to be utilised in a comprehensively planned leap to create "a just and prosperous society of Indonesian socialism."

Furthermore, Kuntjorojakti (1978, 141–142) states,

The resemblance of the Guided Democracy central-regional system to the highly centra-l-ized and autocratic structure of the colonial regime was extremely close. Both started from the same premise of strengthening central political control over the regions and both pursued almost the same strategy of arranging the organic relation between the higher and lower local authorities and in deciding the method by which the centre delegated specified authorities or functions to the lower level.

The change in the national political constellation since "guided democracy" in which the military had taken a decisive role in directing government policy has significantly influenced the subsequent course of transmigration policy. The ultimate goal of transmigration was stated to increase security and improve welfare through the attainment of three intermediate goals: the opening up of new areas of natural resources and land in a proper manner; moving the population from densely populated areas to empty areas; and developing strategic regions in order to achieve greater national resilience under a just and prosperous Indonesian socialist society (Hardjosudarmo 1965, 129–130). In 1961, another ambitious transmigration plan was presented as part of the so-called eight-year development plan (1961–1968) formulated by the historian, poet, and politician, Professor Muhammad Yamin.[12] In the plan, the transmigration goal was described as being "to move at least 250,000 transmigrants to the agricultural sectors in the Outer-Islands" (Heeren 1979, 23). According to Heeren (1979, 23), there is confusion among some sources about whether the target number of the plan refers to persons or families. On the plan itself, Heeren (1979, 24) also commented, "*It is obvious that Yamin in fact gives little attention to transmigration. In the plan, transmigration policy, for example, was*

[12]Muhammad Yamin was described by Anderson (1972a, b, 288) as "Indonesia's best known political eccentric as long as he lived. It was said of him, not altogether unkindly, that he was like a horse: if you were in front of him, you were likely to get nipped; if you were behind him, to get kicked; and if under him, to get trampled on. But if you were on top of him, with reins in your hands, he would carry you fast and far."

mentioned under the rubric of Government and Police! The budget for transmigration is only 1.5 percent of the total government budget."

The government's intention to link transmigration policy with "Government and Police," however, is not surprising since the military played an important role in the ruling elite of the time, as well as within the bureaucracy which formulated the plan. As might have been expected, the actual number of transmigrants moved during the Yamin Plan was very much below the target. Between 1961 and 1965, only 92,473 transmigrants were moved. One of the main receiving areas during this period was the district of Siabu in Riau Province. A later study showed that strategic reasons had been partly responsible for this resettlement, in order to support military operations during the confrontation with Malaysia (Hamidy and Ahmad 1984, 10). In 1964, an evaluation of the Yamin Plan and a review of transmigration problems led, in January 1965, to a policy called *Transmigrasi Gaya Baru* (the new transmigration model). It was announced directly by President Sukarno who stated, "*No, Indonesia does not need birth control; Indonesia can feed at least 250 million people. 250 million people, if they are properly distributed, if their energies are utilised adequately.*"[13] Under the new plan, the government aimed to move 100,000 families every year. The record shows that in the first semester of 1965 the government successfully moved almost 25,000 transmigrants (Heeren 1979, 21). As a result, of the political turmoil after the abortive coup of October 1965, however, the implementation of transmigration was significantly impeded.

Transmigration policy as practiced by the Indonesian Government to solve the population and related problems of Java was described by Wertheim (1959, 196) as not only impractical, but also illogical,

> *Why repeat a pattern which has produced in Java a scarcely enviable situation? Why should a solution of the population problems of the outer islands be exclusively sought in the agrarian sector, which inevitably leads to increasing competition for land? Why not try, in this early stage of development, an industrial development which would attract workers both from Java and Sumatra with appreciably less friction.*

Jaspan (1960, 3–4, cited in Hardjosudarmo 1956, 122) argued that transmigration as practiced by the Indonesian government was only "old stuffing with new cloth," and merely a "dead end" (*jalan buntu*). He also wondered why the emigration policy, which essentially served the capitalists' interests and which was characterized by its exploitative character, was continued by the Indonesian government whose ideology, "Pancasila," was obviously opposed to such exploitation (cited in Hardjosudarmo 1965, 124).

[13]Quoted from Sukarno's speech in the general meeting on National Movement for Transmigration on December 28, 1964, entitled *Soal Transmigrasi adalah soal mati-hidup bangsa Indonesia* (transmigration is a matter of life or death for Indonesia) (Departemen Penerangan RI c.1964, 13).

Regional Responses to Sukarno's Transmigration Policies

The resentment of the indigenous population in the settlement areas toward the central government's implementation of the transmigration policy in the 1950s and 1960s was not a new phenomenon. The sources of resentment and dissatisfaction were mostly related to land disputes. Problems arose partly because the concept of land ownership in the Outer Islands was based on customary (*adat*) law, in which all land belonged to the *marga* or clan. The reluctance of indigenous people in Lampung, where most population resettlement was implemented, to grant their land for colonization in the Dutch period, according to Heeren (1979, 47), had resulted in the formation of Javanese enclaves separate from the indigenous settlements. When the local people refused to give up their homeland for colonization, the Dutch used *perintah halus* (gentle commands) to gain permission to use land outside *marga* jurisdiction. The areas for colonization, therefore, were usually located outside the local settlements.

Nitisastro (1955, 125, cited in Heeren 1979, 17) from a somewhat nationalist stance criticized the Dutch assumption that had led to "enclave politics." According to Nitisastro, the result of the typical Dutchman's conservative politics is blockage of the assimilation process among ethnic communities in the colonization areas. Furthermore, Nitisastro argued that enclave politics is obviously in contradiction with the primary goal of Indonesian independence, that is, to build one Indonesian nation. Assimilation of the Javanese within indigenous populations in the Outer Islands, which was expected to bolster the unification of Indonesia, was not as easy as the government or the "nationalists" like Nitisastro had anticipated. Wertheim (1959, 196), based on his observations in Lampung in 1956, noted that.

> *Assimilation of Javanese migrants is proceeding, neither to a Sumatran nor to a general Indonesian society, but to a Javanese society somewhat modified by a Sumatran environment. This situation leads to an increasing resistance of Sumatrans to the way in which resettlement has been carried on. Such resistance may seriously hamper further transmigration efforts. Thus, it can be stated that the absorptive capacity of the outer islands is not only restricted by spatial and technical but by social factors as well.*[14]

In 1950, when the government of the Republic of Indonesia in Jakarta began to implement the transmigration policy in the Outer Islands, the local administration system was also experiencing rapid change. One consequence was that under central government pressure, the land ownership system based on *adat* law was slowly being transformed into new arrangements. Changes in the system of national administration deliberately reduced the role of the interests and views of indigenous populations in the Outer Islands. A reorganization of regional government structure and hierarchy was introduced, and a new autonomous regional government unit

---

[14] Wertheim's prediction on the probability of social conflict in Lampung has apparently been realized in the "Lampung Affair," which broke out recently (see Tempo, February 18, 1989). See also an interesting personal account from Wertheim (1989, 20–21) on this affair in *Inside Indonesia*, July 1989.

called *negeri* replaced the *marga* system. According to this new arrangement, a *negeri* could be formed from one or more *marga* and include the Javanese population living in this area. According to the central government, the organization of a *negeri* was to be based on democratic principles. The council and the head of a *negeri* were to be democratically elected. This new rule was needed to supersede the *adat* law, which failed to grant the right to vote (*hak pilih*) to Javanese migrants. Through the new election system, *negeri* could be dominated by the Javanese migrants in those areas where Javanese outnumbered the indigenous population. This situation understandably created further disappointments with central government in the eyes of local people. Since they could not refuse the imposition of *negeri,* they tried to have separate *negeri* in which they could still practice their *adat* laws. Although in the end they failed, they frequently sent resolutions (*resolusi*) and motions (*mosi*) to the central government through their organizations in Lampung, pressing for more tradition-based reforms (Heeren 1979, 47–48).

A further change in local administration was introduced by the central government in 1959, in conjunction with the declaration of guided democracy by President Sukarno. Liberal democracy was considered inappropriate for Indonesian conditions. Under the new system, the *marga* and *negeri* were abolished. Regional authority was held by the regent (*bupati*). Under the regent, the *camat* (*asisten wedana*) was the most powerful authority in the regions. Both *bupati* and *camat* were appointed by the higher authorities. Such structural arrangements, in turn, contributed to the circumstances in which neither the local people nor the migrants had opportunities to influence the course of regional government.

The central–regional relations are indeed a crucial issue in Indonesia. As Kahin (1963, 674) observed in the early sixties that.

> *Regionalism is still an important force in Indonesia-scotched, but not eliminated, by the suppression of the 1958–1961 rebellion and having today considerable potential strength. The course of the rebellion and its overwhelming defeat demonstrated the strength of Indonesian nationalism to be greater, or at least more enduring, than most outside observers had assumed. But many of the conditions which led to the rebellion remain. The provinces' discontent with Jakarta's paternalism is still profound. Local dissatisfaction with the central government and the belief that decentralization of administrative authority and fiscal power would permit the regions to promote their own welfare more effectively remain strong.*

A specific issue that stimulated strong criticism from the indigenous people, according to Heeren (1979, 48), focused on the government budget, which allocated more to transmigration settlements and less to other areas. Road construction, schools, agricultural offices, health, and many other facilities were usually concentrated in the transmigration areas with smaller allocations to other places. However, regional disappointment with central domination had a much broader base, as argued by Paauw (1963, 156),

> *Resistance among the peoples of the outlying islands to the President's new program for Indonesia's development reflects fear of centralization of authority as well as more moderate attitudes toward capitalism and foreign enterprise. Behind these attitudes we find both ethnic considerations and significant differences in historical experience. The contribution*

of foreign enterprise to local welfare was more real and obvious than on Java, the burden of colonial control less onerous, and the desire to maintain local autonomy more respected. Individualism in economic activity is more highly valued in many of the outlying islands than on Java, producing conflicts over the degree of economic centralization to be promoted by political decision.

In such circumstances, transmigration policy, which was basically aimed at resettling people from Java to the other islands, has further exacerbated the discontent among local people toward the central government. Local discontent with transmigration policy among the indigenous population in South Sumatra was clearly articulated at an *Adat* Congress, firstly in Palembang in January 1957 and later in Bukittinggi in March 1957. The following is the full text of a resolution, which resulted from the last congress in Bukittinggi.[15]

All transmigration (policy) to Sumatra should be temporarily stopped. The reason is the fact that the number of transmigrants has reached 40 percent of Lampung's population. To avoid further disappointments among the indigenous population, transmigrants should be treated equally with other people in Lampung; for example, they should have to follow the customary law wherever they settle. The central government should give attention to the important sectors of the indigenous people which are lagging behind the facilities for transmigrants. The areas that are still unoccupied by transmigrants have to be allocated for the local people who move as local transmigrants. The central government has to implement local transmigration as well as general transmigration, and the budget should be provided for it. The general rules for transmigration policy have to be re-evaluated, and the new rules should be based on the interests of the regions. The areas for transmigration also have to be selected based on regional opinions. The decision to move transmigrants, as well as the number of transmigrants selected, should be agreed to by the regions. The regions also have the right to determine the requirements needed for transmigrants who will be sent to them. The transmigration officials in the receiving areas should be people from the region, or at least appointed by agreement with the local people (Heeren 1979, *49–50, author's translation).*

Heeren (1979, 50) also noted that the abandonment of transmigration policy was one of the main demands from the PRRI rebel groups (Dewan Banteng in West Sumatra and Dewan Garuda in Palembang) to the central government between 1956 and 1958. The implementation of transmigration was suspended during these years of rebellion but resumed in April 1958 immediately after the PRRI was crushed.

Summary

The colonial state's ethical policy at the turn of this century affected the course of Indonesia's development. One of the enduring results was the migration policy, which relied on sponsored population mobility from inner (namely Java) to Outer Islands. This policy has remained on the state agenda for almost a century. In a much broader sense, the dichotomy between "inner" and "outer" islands, conceived by the

---

[15]This resolution appeared in *Indonesia Raya* daily newspaper, March 20, 1957.

colonial state, has also persisted. This dichotomy apparently resulted from the perception of the Outer Islands as empty land and the main sources of economic revenue and of Java, as the center of political power, and plagued by a population surplus. The persistence of a population resettlement solution, rather than industrialization, reflects the continuation of the dominant interest within the state, which serves the elite of the day. Migration policies, no doubt, are merely palliative in nature, and consciously avoid any fundamental solutions, which demanded redistribution of productive land or capital. Although in the early years of independence it was assumed that transmigration would be integrated with industrialization, such hopes were soon thwarted.

Although the early years of independence produced an enthusiastic supposition that transmigration would be integrated with industrialization, such hopes were soon thwarted. This was partly due to the unsettled political situation and the unstable bureaucracy, as well as to the limited government funds. National leaders were mostly absorbed by the struggle to gain and hold national sovereignty, particularly against the former colonial power, as well as to unite the country in the face of the centrifugal power of the regions. Within such countervailing pressures, particularly between the regions and the center, the old outer–inner dichotomy became the dominant issue of the first two decades of independence. It is within those central–regional contexts that transmigration policy maintained its instrumental value. The colonial legacy of "*emigratie*" as a tool of welfare policy dominated the course of transmigration policy during the first decade of independence. It promised the government a solution to population and related problems in Java. These included relief of poverty and landlessness and help for the victims of the Islamic rebellion in West Java, repatriates from Surinam, and ex-soldiers and guerilla fighters. The resentment of indigenous people in the Outer Islands such as in Lampung, toward the central government policy on transmigration, mostly originated from a different conception of land ownership. The people in many parts of the Outer Islands still regarded their lands as being inalienably owned by their customary groups (*marga*), whereas the central government considered that the national interest should prevail over regional interests. The regional disappointments with transmigration were openly articulated whenever there was a channel to do so. The regional rebellions against the central government between 1956 and 1958 provided the most dramatic occasions on which local people in the Outer Islands expressed their resentment toward the transmigration policy. Strong resentments among people in the Outer Islands suggest that the aim of providing a "melting pot" for Javanese and non-Javanese through transmigration was not without great risk.

Partly as the result of regional rebellions in West Sumatra and South and North Sulawesi during the period 1956–1958, transmigration gained a new aim as a strategic instrument to strengthen national integration and security. The increasing role of military leaders in the government bureaucracy played an important part in establishing the strategic aims of transmigration. Up to the time of the attempted coup of 1965, although the total number of people who were moved under these schemes was small, some resettlement areas were designed to bolster the national defense, particularly during the dispute with Malaysia. In the 1950s and 1960s, transmigration was always ambitiously planned, but its implementation was consistently poor. Expertise and funds were chronically limited. However, the long-lasting perceptions that have valued transmigration as a multipurpose instrument may also have mystified transmigration as a "panacea," a cure for many diseases. In the New Order period that began after 1965, such perceptions have apparently been maintained. This helps to explain many controversies about the implementation of transmigration policy during the New Order.

While the Dutch colonial government used emigration policy only implicitly as an instrument of demographic engineering for security purposes, after independence, the government of the Republic of Indonesia explicitly relocated people for political and strategic goals. Demographic engineering has become an important part of the state's strategic policy to reduce the increasing possibility of social and political tensions, particularly in areas surrounding the national capital city of Jakarta. Transmigration policy has also been explicitly utilized to support the state's foreign policy during confrontation with Malaysia, among other strategies by relocating people to the border areas, such as Riau, and West and East Kalimantan. After independence, several ideological leaders, such as Yamin and Sukarno, also perceived transmigration to be a strategy aimed at assimilating different ethnic groups to build a homogenized cultural identity and national integration.

Chapter 2
Transmigration and New Order's Development Plan

> While the policy maker may not make the values or interests underlying his policies explicit—indeed, it may be in his interests not to—it is the task of the policy analyst to do so.
> *Myron Weiner (1975, 76)*

National Development Policy: An Overview

During the period of the New Order, economic development had become a major concern of the ruling elites in Indonesia. The emergence of a group of economists dedicated to the promotion of economic growth has been clearly reflected in the establishment of national and regional five-year development plans. These plans stress rational economic calculation as the primary consideration in promoting economic growth, and in this context, population is regarded as only one component of the economic system. The conventional view among these economist–technocrats is that the rate of population growth is a significant determinant of the success of a country's economic development efforts. The importance of the population factor in economic thinking is clearly demonstrated using average annual per capita income as a measurement of development. The effect of population growth on economic development has been explained, by Coale and Hoover (1958, 19), among others, who argue that a higher rate of population growth implies a higher level of needed investment to achieve a given per capita output, while there is nothing about faster growth that generates a greater supply of investible resources. As a consequence of such an economic preoccupation, the reduction in population growth is often considered to be a necessary precondition for economic development.

The legacy of colonial policies, as well as the population policies of the "Old Order" (which were basically pronatalist and viewed uneven distribution of population as the main population problem), could not be easily eliminated from the thinking of the ruling elites. An illustration of how these elites, particularly the military, have persistently regarded transmigration as an important undertaking is shown by a request from General Suharto to the US ambassador in Indonesia, Marshall Green, at their first meeting on May 29, 1966. Suharto asked for US$500

R. Tirtosudarmo, *From Colonization to Nation-State*,
https://doi.org/10.1007/978-981-16-6437-3_2

million in grants or soft loans to assist the transmigration program.[1] Compared with the problem of rapid population growth, which at the beginning of the New Order was accorded high priority, the problem of uneven population distribution between Java and the Outer Islands has not been generally regarded as an urgent issue by the economist–technocrats. Widjojo Nitisastro (1970, 238), widely known as the architect of the New Order's economic development policies, strongly argued that what was needed to overcome the population problem in Indonesia was "*a massive development effort to create expanding employment opportunities accompanied by a rapid spread of fertility control.*"

Yet, curiously enough, the problem of uneven population distribution and the role of transmigration as a means to overcome it have long been a focus of government thinking and became a very important policy within the national development plans during the New Order. The real motive of President Suharto for the continuation of transmigration is not difficult to identify. The explanation might be sought in the idea of "harmony" among the Javanese, which in the Indonesian political context can be translated into the concept of "national unity" or "national integration," and which, as suggested by Koentjaraningrat (Visser 1988, 75), is strongly represented by the president and the military. In this context, transmigration has been perceived as an instrument to accelerate the process of national integration.[2] Government policies to move people from overcrowded Java to the other islands have also been considered as an alternative to land reform. Land reform, which was aggressively promoted by the communist party (PKI) prior to the abortive coup of 1965, was identified by the New Order as a "communist" policy. Wertheim (1984, 10) noted, "*Under the Suharto regime, 'land reform' has become, for over one-and-a-half decades, a forbidden topic because it has the 'bad' smell of 'Marxism' and 'Communism'.*" Tjondronegoro (1972, 12–13) noted that the shift in the balance of political power to the military after 1966 resulted in a significant policy change from land redistribution within Java to population redistribution across the archipelago. Tjondronegoro (1984, 79) explained that.

> *Since 1967, the government has given higher priority to economic production and technological improvement, carefully avoiding policies which could have brought about effective social reforms. The Basic Agrarian Law of 1960 which is a potential instrument for initiating structural changes in the rural areas has not been effectively enforced since the Coup of 1965. Since then, perhaps, the only occasion on which the government did implement this Law with regard to rural structural change, was when it had to set the land allocations for the migrants in the Outer Islands.*

These seemingly ambiguous perceptions among the Indonesian elites of the nature of population problems are not unique. Weiner (1975, 66) has suggested that many governments assume that some of the problems exemplified by rapid

---

[1] A letter from Marshall Green to Dr. Terry Hull at ANU, March 18, 1988.

[2] In more recent times, Martono, then Minister of Transmigration, stated clearly that the ultimate goal of transmigration was a new society resulting from the amalgamation of different ethnic groups (Hopper c.1987, 197).

population growth can be alleviated by policies aimed at the dispersal of populations. He also noted that migration policies are frequently embraced for a variety of purposes, but almost never for exclusively demographic reasons. In addition, transmigration as a form of land settlement scheme can be a supreme rhetorical device for the government and its ruling elites. Hulme (1987, 413) observed that "*As a panacea for economic and social problems, settlement schemes are a strategy almost without rival. If the popular expectation is that those in power should 'do something', then schemes are a most effective means of being seen to 'do something'.*" Analyzing transmigration policy within the context of national development plans requires it to be assessed, in addition to the demographic perspective, on the basis of two other influential considerations: first, economic considerations, as proposed by the economist–technocrat group and the World Bank; and second, noneconomic considerations as advocated by the military, among others, particularly the social and political aspects of national integration. Alternating states of conflict and consensus between the groups promoting these two perspectives have played a decisive role behind the scenes in the formulation of transmigration policy and, more importantly, during its implementation. Preceding the launch of the first five-year plan in April 1969, the government, under the supervision of the economists, successfully reduced the inflation rate from a high of more than 600% in 1966 to around 10% in 1969. This achievement indicated the important role of the economist–technocrats in effectively controlling government expenditure and the money supply.

In contrast to the situation under the Old Order, after 1967, the government's budget had to be submitted to parliament before the start of each new fiscal year. By 1968, it had become obvious that the central government had regained control of the economic situation. This was considered a suitable time to draft the first five-year development plan. As explained by one of the New Order's leading economists, the first five-year plan was not an exhaustive plan of expenditure and was not grounded on any particular macroeconomic model (Mangkusuwondo 1970). Mangkusuwondo (1970, 17–18) explained that:

> *It does not start with aggregate targets such as the level of GNP, total consumption, investments, exports, imports,* etc. *It does mention a target rate of growth of total income of five percent* per annum, *but that figure is nothing more than a guess about the rate of growth of income as a consequence of investments in several areas of the economy, since neither the level of income nor total investment is known with a reasonable degree of confidence.*

Based on their knowledge of conditions in the Indonesian economy prior to the beginning of the first five-year plan, the designers decided that it was more realistic to frame a sectoral plan, rather than a comprehensive one. The major emphasis was placed on the agricultural sector, notably rice and other food production, and on the export sector. Other sectors, such as social overheads and manufacturing industries, were emphasized only insofar as they contributed to development in the food and export producing sectors or led to foreign exchange savings.

Since the first plan, a sectoral framework has always been used by the government as a basis for national development programs.[3] During the preparation of the first plan, the Indonesian government received a great deal of financial and expert assistance from the World Bank, particularly in the formulation of its economic policies and programs. According to Thompson and Manning (1974, 56), the involvement of the World Bank in Indonesia began in 1968 when Robert McNamara visited Indonesia as president of the bank. Up to 1968, the role of the World Bank had been minimal. Although Indonesia had become a bank member in April 1954, because of the political situation during the mid-1950s, the World Bank did not consider it economically feasible for the Indonesian government to borrow bank money. A report based on the World Bank's first field mission to Indonesia in mid-1955 suggested that "*The bank should take no decisions on further operations in Indonesia until the political situation is clarified and the ability of any future government to tackle some of the critical and financial problems can be more clearly assessed.*"[4]

At the beginning of the New Order period, the World Bank staff worked directly with the National Development Planning Board (*Bappenas*), the Ministries of Trade and Finance, and the Foreign Investment Board (Thompson and Manning 1974, 59–60). The central position of *Bappenas* within the national policy was explained by Rudner (1976, 255):

> *As a central planning institution Bappenas had substantive responsibility for the formulation of development strategies expressed in its sectoral and overall plans, consistent, of course, with politically—determined modernization goals. In the event, Bappenas also acquired a pivotal function in current development policy decisions. It was responsible for getting annual departmental budgetary proposals, and in close collaboration with the Finance Ministry, aggregating these into the state development budget. Foreign project aid, as well, was disbursed through Bappenas. The Planning Board coordinated project aid finance and even conducted the negotiations with foreign governments and multilateral and international organizations.*

At the end of the first five-year development plan, the People's Consultative Assembly (MPR), the highest constitutional power in Indonesia, promulgated, for the first time, a policy guidance document called "the broad outlines of state policy" (GBHN). Based on the GBHN, which provided the basic strategy for the national development policy, the subsequent five-year development plan (second plan) was formulated. In theory, each five-year development plan is an operationalization of the GBHN. According to Emil Salim, one of the New Order's economists (1970, 1, quoted in Djamin 1984, 23): "*Repelita is merely an indicative plan, which shows the direction in which available resources will be focused. It provides information about factors that could hinder the dynamics and the speed of the development process. In addition, Repelita also reflects a style and system of economy that has been adopted by the government*" (author's translation). Thus, the statement of a

---

[3] It is interesting that the original framework is still being used by Bappenas, even though the data to develop more comprehensive plans are readily available.

[4] IBRD (1956 cited in Thompson and Manning 1974, 57).

numerical target (e.g., the number of transmigrants that were to be moved, the number of schools to be built, or the number of family planning acceptors to be achieved) in the *Repelita* (five-year development plan) documents was only an indicative goal. The actual numerical target was proposed by the planners on the basis of each department's projected budget, as well as the expected capacity of the bureaucracy, which played a crucial role in the implementation of the plan. Consequently, the numerical target for transmigration has rarely been calculated in relation to other demographic indicators, such as the rate of population growth or the rate of interprovincial migration.[5]

The five-year development plan is further elaborated in an annual plan that shows, among other things, the priority of each sector, the budget ceiling, and the sources of finance for each sector. The annual plan is contained in the national budget (APBN). A new rationalized budgetary process, based on a fiscal year extending from April to March of the following year, was initiated at the beginning of the first plan. Mas'oed (1983, 239) noted that in the preparation of the APBN at the national level, the finance minister and the head of *Bappenas* have acted as the "Guardians of the Public Treasury" and the other departments and agencies as the "Spending Advocates." The national policymaking process under the New Order has been described by some observers as predominantly technocratic–bureaucratic in nature (Kuntjorojakti 1978; Mas'oed 1983). It is bureaucratic because it always originates in, and is implemented by, the bureaucracy. The parliament functions mainly to ratify the plans, and debate on the matter is not encouraged. It is technocratic because the main criteria used in the process are technocratic values, such as "rationality" and "efficiency": The economists are anxious to implement such criteria in evaluating project proposals and carrying out feasibility studies. In a similar vein, Hugo et al. Jones (1987, 300) argued that the politico-planning structure in Indonesia can be summarized in three words: "authoritarian," "centralized," and "technocratic." Glassburner (1978, 29) views the planning process and the role of economist–technocrats in a slightly different way:

> *Although Bappenas is the centre of activity at the time of writing of the Five-Year Plans, the agency itself is not set up for highly sophisticated central management of the economic system. Its primary function is to consider petitions from government agencies at various levels for investment funds channeled through the budget. I believe it is much more accurate to view this process as a manifestation of aristocratic benevolence than as an Indonesian version of Gosplan [Soviet Central Planning Office].*

---

[5]The current Minister for Transmigration, Soegiarto, in interviews with the press, stated clearly that the Repelita target number for transmigration was merely prepared for bureaucratic reasons, particularly in association with the proposed budget allocation (see Suara Pembaharuan, April 25, 1988, also Editor, May 7, 1988).

The Role of Provincial Government

Social, political, and economic structural changes since the end of the 1960s have contributed to the development of the centralized character of the Indonesian government. In 1974, for example, the New Order government passed Law No. 5, which established the legal framework for the role and authority of the provincial governments (Handajaningrat and Hindratmo 1986, 5). According to this law, the provincial government authorities exercise only those powers that have been granted to them by the central government. The national government is primarily represented by the Department of Home Affairs, whose minister is granted extensive authority under the law. With the establishment of this authority, the Minister of Home Affairs, acting through the governors, has acquired responsibility for supervising and supporting all development activities carried out in the provinces. However, in practice, there seems to be a dualism of authority at the provincial level, since most regional development programs have been implemented by the national sectoral line agencies (Departments and Boards). The role of the departmental regional offices *(Kanwil)*, as representatives of the national government departments, is not clearly distinguished from the role of the district technical offices (*Dinas*), which fall directly under the control of the governor or the district head (*Bupati*). The fact that most of the development budget is channeled through *Kanwil* offices for sectoral programs inevitably means that national priority programs tend to take precedence over regional ones.

Development activities at the provincial level, based on Law No. 5 of 1974, can be classified into three broad categories. First are the development activities and programs implemented by the sectoral line agencies in the regions. In this category, planning, implementation, control, and financial support are mainly the responsibility of the central government. The transmigration program is an example of such an activity. Second are the development activities funded and implemented by the district technical offices *(Dinas)*. Third are the development activities that should be the responsibility of a provincial government, but which, because the financial ability of the provincial government is limited, usually become the responsibility of the central government. An example of such activities is the *Inpres* projects that are implemented by the provincial government agencies under the strict control of the central government, particularly the Ministry of Home Affairs[6] (see Fig. 2.1). In 1976, the government created Regional Development Planning Boards (*Bappeda*) to strengthen the regional government in planning and co-coordinating development programs. In theory, there are a number of functions that should be performed by *Bappeda*: developing the basic pattern for regional development; preparing the regional five-year development plan; proposing such annual programs for the implementation of the five-year development plan; coordinating planning between the

[6] *Inpres* stands for *Instruksi President* or presidential instruction. At present, there are eight different types of *Inpres* grants, one for provinces, one for regencies, one for villages, and the remainder for specific purposes such as health, primary education, roads, and markets.

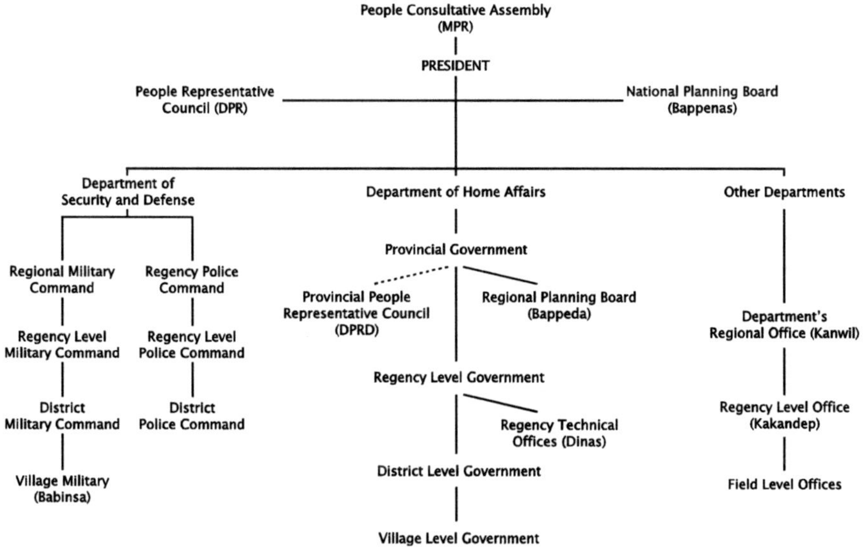

Fig. 2.1 Structure of government

various *Dinas* and offices under the provincial government as well as sectoral agencies; proposing the regional budget; and conducting research and monitoring the implementation of regional development. The structure of *Bappeda* consists of the head, vice head, secretariat, and four divisions, namely research, economy, material and infrastructure, and statistics and reporting. In practice, however, these bodies have had limited control over development programs in their regions. Morfit (1984, 61) noted that, although the *Bappeda*, as planning institutions, is required to produce regional five-year development plans (*Repelita Daerah*), these often are little more than a compilation of data about a given region. The weakness of *Bappeda* is typified by their lack of ability to influence development budgets, which are largely controlled by sectoral agencies.[7]

Given that sectoral line agencies in the provinces are directly controlled by their national offices in Jakarta, the governor has a duty to back the decisions of sectoral line agencies for the region. Bappeda, which in theory has the authority to evaluate the implementation of all projects in the region, is entirely peripheral, since it is

[7]Several attempts, initiated by the economist–technocrat group in *Bappenas* and Home Affairs Department, have been made to strengthen the role of *Bappeda* as the provincial planning institution. For instance, before the preparation of the annual budget for the provincial level, *Bappenas* has conducted national-level workshops in which all the *Bappeda* representatives participated. On these occasions, the *Bappeda* staff were briefed and guided to prepare their own annual regional budgets. Through *Bappenas*, several studies, sometimes initiated by government departments and very often involving foreign experts, have also been conducted to assess the possibilities in strengthening the planning capacity and roles of *Bappeda* in regional development (MacAndrews et al. 1982; van den Ham 1989).

subservient to the governor. The total dependence of the governor on central government decisions was noted by Major General Eddy Sabara, a former Inspector General of the Department of Home Affairs, and the agency in charge of selecting new governors. Sabara (Editor, September 26, 1987) explained that, at the provincial level, the selection of the governor was nominally under the control of the regional military command. Thus, the nominee for governor must be a candidate who is considered politically acceptable by the local military command. However, they would ultimately refer the matter to their superiors in Jakarta. Santoso (1986, 51) also noted that "*He [the governor] is more a central government organ in the region than a representative of the region itself.*" The circumstances in which the governor and the governor's office apparatus function solely as an instrument of the central government have undoubtedly contributed to the "top-down" and "centrally planned" regional development approach.

The First Plan, 1969–1973/1974

Between the political upheaval of 1965 and the commencement of the first plan, the government paid little attention to transmigration. At the dawn of the New Order, McNicoll (1968, 69) even predicted that the program would no longer occupy a significant role in government policy. Heeren (1979, 171), after studying the various forms of migration schemes, argued in 1967 that the future trend of migration in Indonesia would be toward voluntary internal migration. These predictions were based partly on an assessment that the economic and political environment did not favor such a policy. The economic and political situation, although under government control, was still considered unstable and very fragile. In the first plan, therefore, stabilization and rehabilitation became the immediate targets. The technocrats, who were mostly economists from the University of Indonesia with backing from the military, formulated a relatively modest plan with emphasis first on increasing food production, and second, on rehabilitating the physical infrastructure, such as roads and ports. Although the economist–technocrats seemed to dominate the field of advice and expertise, various centers of economic power were essentially controlled by military personnel. Glassburner (1978, 34) observed that:

> *The problem of economic-policy making, and implementation is a major problem of social engineering in a pluralistic milieu. The FEUI [Fakultas Ekonomi Universitas Indonesia] economists are thus forced to rely on their expertise and their persuasiveness, as well as their ability to appeal to the national interest. Even within the ministries they head they must share power and responsibility in important instances.*

The dominant role of the military in the formulation of policy was described as follows by Rudner (1976, 250):

> *The new development strategy sought to improve the performance of the laggard Indonesian economy not merely for its own sake, but as a means of securing the military elite's political aims of social modernization with stability, and national resilience. As the vanguard of the New Order, the ABRI, and particularly its army component, was to play a critical role in the*

formulation and implementation of the First Five Year Development Plan (Repelita I), introduced in March 1969.

Prior to the initiation of the first plan, the Directorate General of Transmigration was converted into the Department of Transmigration and Cooperatives as a result of a merger between the Directorate General of Transmigration (under the old Department of Transmigration and Veterans Affairs) and the Directorate General of Cooperatives (formerly under the Department of Home Affairs). In March 1969, Brigadier General Soebiantoro was appointed Director General of Transmigration, replacing Brigadier General Boesiri. The appointment of military officers within the government bureaucracy, although very common under the New Order, had a special significance in the transmigration program because it provided an avenue for the armed forces to pursue their strategic and political interests in various aspects of the program.

In the first plan, the government's policy on transmigration was stated in only two-and-one-half pages (Departemen Penerangan RI 1969, 537–539). Although the problem of population redistribution was mentioned, it was not elaborated upon. The need to meet labor requirements for development in the regions outside Java, although stated as a secondary goal, was given more detailed attention in the plan. It seems that the economist–technocratic views that considered transmigration primarily an instrument to support the immediate objectives of stabilization and rehabilitation strongly influenced the formulation of the plan. The transmigration program was considered as an important source of labor to serve these broader goals. In addition, the transmigration program was expected to increase food production in the Outer Islands. In the first plan, no numerical targets for transmigration were set. This was probably because of difficulties in obtaining reliable aggregate data, a major problem for the planners who prepared that document. In addition, the focus of transmigration seemed to be more on regional development than on solving the problem of uneven population distribution between Java and the other islands. Among the economist–technocrat group, the view that transmigration policy should be directed toward regional development rather than to solve the demographic problem in Java was not a surprise. In 1963, for example, Sadli (1963, 22) argued that "*transmigration has to be seen more as a way of providing manpower for economic development in the outer regions.*" In commenting on this new perception of transmigration, Arndt (1984, 6) noted that:

> *As rational economic policy-making again became possible after the 1965–66 change of government, interest in transmigration revived, but with a different perspective. It came to be recognized that transmigration could not solve, or even substantially alleviate, the problem of "population imbalance" between Java-Bali and the other islands, if indeed that was the problem. Transmigration was now seen primarily as having a welfare objective, to raise the living standards of the migrants and that of their home villages in Java. It was also hoped that transmigration could serve as a strategy for regional economic development in the outer islands.*

The declining role of transmigration policy as a solution to population problems was also noted by Schultz (1972, 117): "*The (First) Five Year Plan, for the first time in Indonesian history, not only establishes population control as a legitimate objective of development planning but also makes the realistic assessment that transmigration, or the transporting of Javanese and Balinese to the outer islands, is not an effective means of population control.*" The actual course of transmigration, however, was neither as predicted by many observers nor as stated in the plan documents. In 1970, under the patronage of Brigadier General Soebiantoro, a national seminar on transmigration was held in Cipayung to discuss the role of transmigration in national development. The seminar, which was attended by scholars and policymakers, was a landmark because it restored transmigration to a position of national importance and formulated a positive basis for the transmigration program within the framework of national development policy. The document, published as a seminar report, shows that uneven population distribution was detrimental to many aspects of national development and, more crucially, to the achievement of national integration and security (Seminar Transmigrasi 1970). The seminar report reflected a strong perception within the government elite, as well as among scholars, that once the demographic goals of transmigration were realized, that is a balanced population distribution, other goals, such as economic, social, and national integration, would be automatically fulfilled.

The issues of "national integration" and "national unity" in fact had been discussed from the early years of independence, and their importance reached new heights during the 1956–1958 regional rebellions. The decisive role of the military in ending the rebellions was a major factor in reshaping the role that military leadership has played in the national political scene since that time (Feith and Lev 1963, 37). In addition, "national integration" has become the key issue and prime concern for the military leaders, whose power was further entrenched by the political chaos of 1965. The conceptual links between transmigration policy and national integration were explained by Brigadier General Edi Sugardo (1986) of the National Defense Institute (*Lemhanas*), the "think tank" of the Department of Defense, in the following terms:

> *From the "geo-politic" and "geo-strategic" point of view, transmigration policy is an effort, method and means to achieve "geographic integration," which is a basic condition for the unity of the nation, to improve the people's welfare and awareness, and provide an opportunity to utilize natural resources. Achievement of geographic integration, on the one hand, will protect the nation from centrifugal tendencies, and on the other hand, strengthen centripetal forces. Geographic integration is also a medium to build understanding, friendship, and co-operation with other countries. From the point of view of national defense, transmigration is supportive of the "Popular Defense and Security System" (SISHANKAMRATA), in order to resist the threat of subversion, infiltration and invasion from abroad, particularly through the provision of human resources.*[8]

Thus, in the view of the military, geographic integration, in which transmigration policy plays an important role by providing human resources, is a necessary condition for the attainment of national integration. Another important element of this military thinking is the concept of "basic defense." The basic defense of the Indonesian armed forces, according to Crouch (1986, 3), was essentially based on the doctrine of "Territorial Warfare":

[8] Sugardo (1986, 4–5; author's translation).

> *This doctrine required that the defense forces, especially the army, be organized principally along "territorial" rather than "functional" lines. The whole nation was divided into territorial commands more or less parallel to the civilian administration. In terms of defense doctrine, a major function of the territorial commands was to maintain contact with the local people so that they could be mobilized quickly to support guerrilla operations whenever necessary. It may be added, of course, that this territorial structure has also been very convenient for the purpose of political control.*

The role of transmigration in the concept of "basic defense" was to populate the empty regions to support the territorial commands.

In 1971, an international-level workshop on transmigration was conducted in which there was active participation by foreign experts. The following year, in 1972, the president established a law on transmigration, which provided the basic legislation for the transmigration policy (Department of Manpower and Transmigration 1979, Appendix 3, 4). According to this law, Indonesian transmigration policy has seven goals: (1) improvements in living standards, (2) regional development, (3) a balanced population distribution, (4) equitably distributed development throughout Indonesia, (5) utilization of natural and human resources, (6) national union and unity, and (7) strengthening national defense and security. The numerous goals set down for transmigration reflected the complexity of the role of the program. These goals also suggested that transmigration tended to be seen by policymakers as a universal panacea, a cure for a wide range of social and economic weaknesses. In 1973, two further seminars were held involving several foreign agencies, including the Food and Agriculture Organization (FAO), the United States Agency for International Development (USAID), and the World Food Program (WFP), which were interested in providing financial assistance for the transmigration program. In the first plan, FAO consultants worked closely within the Directorate General of Transmigration and played an important role in conceptualizing the linkage between the transmigration program and regional development, especially through the expansion of agriculture. The active role of the FAO during the first plan was not surprising given the fact that increasing food production was one of the stated national goals at this time. In one of the published seminar reports, for instance, an FAO consultant, van der Goot (1974, 50–51), suggested to embark on a substantial change in the transmigration program, which at the time was planned to take place in 1980: "*The Indonesian government has to establish and demonstrate its ability to integrate large-scale transmigration and regional development initiatives into a viable and long-term program.*" In the course of time, however, the actual influence of the FAO in transmigration policy formulation seemed to decline, whereas the role of other foreign donor agencies, particularly the World Bank, was later to become significant in reshaping transmigration policy.

The Second Plan, 1974–1979/1980

During the second five-year plan, the transmigration program became part of the Department of Manpower, Transmigration, and Cooperatives. Professor Subroto, an economist from the University of Indonesia, was appointed minister. Soebiantoro, who had been promoted to the rank of major general, continued as director general. In this plan, the transmigration program gained more attention than before. The increase in government revenues from the oil boom provided funds to support the program on a much larger scale than had previously been possible. For the first time, it was stated clearly that transmigration was an important means for strengthening national integration. By contrast, the demographic rationale that had regarded transmigration as a tool for population redistribution was mentioned only implicitly (Departemen Penerangan RI 1974, 451–458). The second five-year development plan criticized the implementation of the transmigration program during the first plan period. The problems identified as hindering effective policy implementation were those relating to settlement preparation in the receiving regions, the choice of location, the provision of equipment, and coordination with the regional government, as well as with other agencies. Prior to the second plan, there were some hopes, particularly among economic observers, that the plan would be prepared in a more systematic way, based on research and the cooperation of many economic experts. But, as Glassburner (1978, 167) argued, because of the rice crisis and the radical impact of changing oil prices meant that Professor Widjojo, the head of *Bappenas,* and his subordinates could not prepare a comprehensive plan. The plan, according to Glassburner (1978, 167), was "no more than a statement of qualitative objectives."

Although the demographic rationale for transmigration was only implicitly stated in the plan, a numerical target was explicitly mentioned. The target was to move 250,000 families within 5 years. The Minister of Manpower and Transmigration, Subroto (1974, 28), argued that the resettlement of large numbers of people from Java, Bali, and Lombok would make a positive contribution to regional development in the Outer Islands. In selecting the sending areas, priority was to be given to regions that were considered "critical," such as areas prone to flooding, and to rural places with population densities of more than 1000 persons per square kilometers. South Sumatra; South and East Kalimantan; and South, Central, and Southeast Sulawesi were designated as the main receiving areas for transmigration settlement (Departemen Penerangan RI 1974, 456). In 1976, a significant development in the transmigration program was begun through the approval of the first World Bank-assisted transmigration project. The World Bank project consisted of two pilot schemes in Southern Sumatra. The first was the rehabilitation of an existing settlement in Way Abung and initiation of a new settlement in Baturaja, and the second was to conduct research on cropping systems and long-term monitoring and evaluation. In the first project, new settlers in Baturaja were provided with 5.0 hectares farms, which included 1.25 hectares cleared for food crops and 1.0 hectares of rubber trees established for the smallholders by a public sector estate (PTP X). These

Table 2.1 Realization of the transmigration program in the third plan (families)

Fiscal year	Target	Realization		Percentage
		World Bank	Hardjono	
1979–1980	50,000	23,078 (1985)	49,772 (24,298)	99
1980–1981	75,000	76,562 (3359)	108,081 (34,428)	144
1981–1982	100,000	93,437 (8961)	123,846 (35,780)	124
1982–1983	125,000	125,269 (32,445)	167,477 (47,124)	134
1983–1984	150,000	217,128 (122,747)	86,298 (24,867)	58
Total	500,000	535,474 (169,497)	535,474 (169,497)	107

Sources: World Bank (1988, 11); Hardjono (1986, 29)

Notes

The number in parentheses represents unassisted or partly assisted transmigrants

The percentage figures are calculated from Hardjono's column

benefits were provided to the settlers as a grant. To promote food crop development, farmers were given cattle for draft power and free agricultural inputs (seed, fertilizer, and pesticides) for 3 years. Transmigrants also received a house and 1 year of subsistence supplies. The project was implemented by a project management unit (PMU) with considerable financial autonomy, and both domestic technical assistance and foreign technical assistance were provided to support the PMU. The first project was considered by the World Bank in its final report (1988, 161) to have been successful.

The second bank-assisted transmigration project, approved in 1979, was intended to resettle 30,000 families along the new Trans-Sumatra Highway, particularly in Jambi Province. The project was also promoted as a model for the third plan (World Bank 1988, 161). The World Bank (1988, 159) reported that bank-supported projects had introduced a few innovative features into the transmigration policy in Indonesia. These included improvements in site screening and evaluation, support for and evaluation of the major farm models, assistance for program management, and development of procedures for utilizing timber from transmigration sites. In addition, the World Bank also admitted that a few problems had been encountered during their involvement in the program. Some of these problems were common to all projects in Indonesia, among them: land allocation, coordination between the implementing agencies, and the problem of quality control (World Bank 1988, 159). However, the bank discovered that some problems resulted from difficulties inherent in programming and implementing a complex sequence of events and from ambitious program targets that had strained the capacity of the implementing agency.

By the end of the second plan, it had become clear that the achievement was far below the target (see Table 2.1). Only 55,083 families had been resettled, compared with the target of 250,000 families. One study indicated that the failure was largely due to the lack of coordination between the various central, provincial, and district offices of the transmigration department assigned to carry out the program (Guinness 1977, 11). Another problem was the limited capacity of transmigration implementing agencies to spend the budget. A special report on "Transmigration Budgetary and Accounting Procedures" by Beddoes (1976 cited in Guinness 1977,

11–12) showed that, in the first year of the second plan, 24% of the transmigration budget had not been spent. In the second year, 34% and, by the second half of the third year, 96% of the budget remained. These represented very large sums, since the transmigration budget had increased from Rp800 million in 1969/1970 to Rp6652 million in 1974/1975 and Rp14,936 million in 1975/1976.

The Third Plan, 1980–1984/1985

In the third five-year development plan, the Directorate General of Transmigration became part of the Department of Manpower and Transmigration, while the Directorate General of Cooperatives was transferred to the Department of Trade. Harun Zain, an economist from the University of Indonesia and a former governor of West Sumatra, was appointed minister of the new department. In addition, a new post of Junior Minister for Transmigration was established, reporting directly to the president but attached to the department. The appointment of Harun Zain, a civilian, as head of the department is solely responsible for the transmigration program, which indicated the interest of the technocrat group in incorporating transmigration into regional economic development. Harun Zain, with his experience as governor in one of the receiving provinces, was expected to play an important role in achieving this goal. However, the appointment of Martono to the position of Junior Minister for Transmigration, a man who was described by Harun Zain as "a political man" close to the president, was also an indication that the president did not want to hand over policy control completely to the technocrats.[9] The division of responsibility between the director general, who reported directly to the minister, and the Junior Minister for Transmigration, who was directly responsible to the president, seemed unclear. Perhaps partly due to this situation but also as a consequence of the very rapid expansion in the size of the transmigration program since the beginning of the second plan, several special decrees were promulgated, particularly Presidential Decree No. 26/1978, in which the organizational responsibilities for the transmigration program were laid down. A series of special bodies—particularly Bakoptrans (*Badan Koordinasi Pelaksanaan Transmigrasi* or Transmigration Implementation Coordinating Body)—were created to coordinate policies, implementation, and administration. Although the numerical achievement of the second plan was far below the target, the third plan surprisingly increased the target to 500,000 households (Departemen Penerangan RI 1978, 304–337). The reason for increasing the target, apart from the euphoria occasioned by the second oil boom, was the economist–technocrats' feeling that the implementation of the program would be improved by the increased support being given by many foreign donor agencies.

[9]The statement describing Martono as "a political man" was made by Harun Zain during an interview with the author, January 1988.

The second World Bank-assisted transmigration project began in 1979, concentrating on the resettlement of transmigrants in dryland areas along the new Trans-Sumatra Highway. Technical assistance was also provided to the Junior Minister for Transmigration to aid in coordination of the overall program. Such assistance was important, given that lack of coordination was considered to have been the major problem in the implementation of transmigration under the second plan. The focus of the second round of World Bank projects on dryland and food–crop transmigration schemes also reflects an important new development. It was a significant departure from the bank's successful first project on tree crop transmigration schemes. The change constituted an important development in the evolution of World Bank involvement on transmigration for at least two reasons. First, the World Bank had failed to influence the policymakers in directing transmigration policy into more market-oriented types of population settlements. Second, although the economist–technocrat group had played a larger role in reshaping transmigration policy into the so-called integrated regional development mode, the ambitious targets of the third plan mirrored the influence of strong forces that continued to view transmigration as a population policy aimed at relieving population pressures in Java through the development of agricultural settlement in the Outer Islands.[10]

The beginning of the third plan, therefore, indicated an important change in transmigration policies. The change could be interpreted, on the one hand, as a revitalization of the noneconomic forces, particularly those represented by the military elite and the government bureaucratic apparatus, and, on the other, as an expression of the difficulties and declining roles of the economic views represented, in particular, by the economist–technocrats and the World Bank. The fact that the World Bank also tolerated the more demographic orientation of transmigration policy suggests that the bank and its bureaucratic elements had adopted a rather pragmatic approach. This, aside from its bureaucratic interest in keeping the project going, also reflected the fact that it had become politically impractical for the bank to withdraw from its extensive involvement in the Indonesian transmigration program.[11] Another factor that also raised the hopes of the government was the success of what it called *Bedol Desa* transmigration. This type of transmigration was carried out in conjunction with the evacuation of an entire population following a government decision to use an area for the construction of a dam or some other large project. The first and apparently the largest project was the relocation of people from some districts in Wonogiri in Central Java to Sitiung in West Sumatra. The success of the Wonogiri–Sitiung project within the *Bedol Desa* program increased confidence among the transmigration planners that the country could carry out similar settlement projects in the future. Wonogiri also happened to be the area of the

[10]The central government obsession with promoting population resettlement based on food crop agricultural systems as practiced in Java, according to Dove (1985, 32), is essentially a manifestation of the agroecological mythology of the Javanese that has developed on Java and strongly influenced the thinking of many policymakers in the central government.

[11]I am particularly indebted here to Geoffrey McNicoll at ANU for raising some points that I have developed further for this argument.

president's childhood home. According to Gondowarsito's study of *Bedol Desa* transmigration (1986, 89–90), it was because of this privileged status that the utmost involvement and cooperation on the part of various government agencies, namely the Departments of Manpower and Transmigration, Public Works, Communication, and Agriculture, were enlisted.

Population distribution and regional development, which had alternately dominated as goals of the first and second plans, received equal emphasis in this third plan. Given the government's focus on equal distribution of welfare, under a policy called "The Eight Paths of Equity" (*Delapan Jalur Pemerataan*), transmigration also became an important vehicle for the achievement of this new goal. The plan also stated that priority would be given to the poorest sending areas, where the problem of landlessness was most acute, and noted that the selection and preparation of receiving areas also needed more careful attention. In the third plan, the government anticipated resettling transmigrants not only in dryland areas, but also in *pasang surut* (tidal) areas. The increased numbers to be settled led to difficulties in providing sufficient suitable land in the receiving areas. Compared to the first and second plans, the third Plan described in detail the facilities and equipment to be provided to transmigrants. The plan seems to have been more seriously and professionally prepared than its predecessors. Coordination among the implementing agencies, which was always at the center of the program's problems, was also given more attention.

With the establishment of *Bakoptrans*, the responsibilities of the various departments involved were more clearly defined. For example, the Department of Public Works was responsible for the selection and planning of sites, for land clearing, and for construction of infrastructure; the Department of Home Affairs was to conduct land-use planning and deal with the legal status of land; the Department of Agriculture was responsible for planning and implementing agricultural development; and finally, the Department of Manpower, Cooperatives, and Transmigration was in charge of the recruitment, transportation, and resettlement of transmigrants, as well as providing them with 5 years of economic and social guidance at each site. Under this highly bureaucratic and hierarchical organizational structure, each department was expected to plan its activities and budget in consultation with the junior minister. However, the authority, budget, and staff of the junior minister's office were far too limited for efficient coordination. Furthermore, coordination proved impossible because each department continued to work independently. In addition, since the Junior Minister for Transmigration reported directly to the president, it was inevitable that hidden rivalry arose between the minister and the junior minister.[12] Under the new arrangement, which was implemented in 1978, the junior minister was given more authority to control the foreign assistance than the minister. This authority considerably reduced the power of the minister to effectively implement the program.

[12] A personal communication with a senior member of staff in the Department of Transmigration, December 1987.

At the end of the third plan, 535,474 families were reportedly have been moved (Table 2.1), although the target of 500,000 was met only by defining 169,947 spontaneous migrants as transmigrants, with or without government assistance. According to Hardjono (1986, 29), to say that the third plan target was reached gives a slightly exaggerated picture of the extent to which people were moved from Java and Bali at government expense:

> *In fact, only 365,977 families were moved by the government transmigration agency as fully supported or general (umum) transmigrants. Furthermore, this figure includes local people who moved into transmigration projects and families that had to be resettled within the same province for some reason. The remaining 169,497 families, representing 32% of the total, were swakarsa (literally, "self-initiative") transmigrants who moved with limited, or in many cases, no government assistance.*

Another observer, Babcock (1986, 182), formerly working as a consultant in the Public Works Department, even described the government's claim that the third plan had achieved its target as virtual manipulation to justify the overly ambitious target.

Widespread skepticism has been expressed about the statistical manipulation in determining program's achievements during the third plan. There are quite remarkable statistical inconsistencies among the several sources that have presented the transmigration realization statistics for the third plan. For example, the figure reported by the World Bank (1988, 11) is obviously different from that presented by Hardjono (1986, 29). Although the total number of resettled migrants is similar, their annual distributions, especially the number of spontaneous migrants, are remarkably dissimilar. While both Hardjono and the World Bank described their sources as the Department of Transmigration, the salient differences in their figures raise serious doubts to the accuracy of the information. Regarding the distribution of spontaneous transmigrants, although the accuracy of both columns must be suspected, the figures in Hardjono's column seem more logical than those of the World Bank. The distribution of spontaneous transmigrants, which were mostly concentrated in the last budget year in the World Bank data, is unlikely to reflect the actual pattern of movement. That is more likely to have been more evenly distributed throughout the plan period, as in Hardjono's figures. The World Bank concentration of spontaneous transmigrants in the last budget year probably reflects hasty decisions by the Department of Transmigration due to pressure at the end of the plan period to document a public image that program targets had been successfully achieved. The method chosen was apparently through manipulation of statistics on spontaneous transmigrants.[13]

---

[13] Such statistical manipulation was recognized by the World Bank, which noted in one of its reports (1988, 134) that:

"Although some spontaneous families were settled in existing areas, no program of partially assisted movement or of support for spontaneous migration was developed in Repelita III. However, the Repelita III target of 500,000 families was eventually achieved by including 169,500 'spontaneous' families in the total moved. Of these, about 30,000 families were moved with partial assistance and the remaining 140,000 moved without assistance but were identified in sending or receiving areas. It is important to emphasize that these families were identified to round out official

The Fourth Plan, 1985–1988/1989

Under Presidential Decree No. 45/M/1983, for the first time in the history of the transmigration program, a separate Department of Transmigration was established. Martono was appointed minister. This ended the previous rivalries between the minister controlling the Directorate General of Transmigration and the Junior Minister for Transmigration. It was hoped that an independent department could solve the problem of coordination, which had been troubled during the implementation of the third plan. The president also established the State Minister for Population and Environment (KLH), who is responsible for the formulation, management, and coordination of national population policies. Professor Emil Salim, a Minangkabau and one of the New Order's leading economists, was appointed as minister. An integrated national population policy had been long overdue. Given the number of economists in the central government, including economist–demographer Professor Widjojo, the lack of integration between population policy and national development policy seemed "paradoxical," at least to outside observers (Hugo et al. 1987, 315). The paradox, however, is hardly surprising given the quite different perceptions of the "population factors" held by the president and his military elite and the economist–technocrats. The difficulties that had been experienced by the economist–technocrat group in dealing with the military were confirmed by one of the Deputy Ministers for Population and Environment.[14]

The political position of the economist–technocrats, as explained by Glassburner (1978, 32–33), was almost entirely dependent on their ability to convince the military leadership, and the president in particular, that their expertise was essential. In a similar vein, Rudner (1976, 255) argued that:

> *Having no political base other than military patronage, technocrat ministers and planners were ultimately dependent upon the confidence of the military elite in general, and the president in particular. Whatever their formal functions in economic policy-making, the decisions of economist-technocrats became operative only when and to the extent that they acquired executive support from military leaders and local commanders. In the last analysis, the authority of the economist-technocrats went only so far as their economics proved amenable to the ABRI's own conception of national modernization.*

In the fourth plan, therefore, it was likely that transmigration would remain largely beyond the influence of the Minister of Population and Environment, Emil Salim, an economist–technocrat, because transmigration had also become an independent department with a minister who was close to the president. Following the decision to make transmigration an independent department, two Directorates of the Department of Public Works were absorbed into the Department of Transmigration. Some sections of the former Directorate of City and Regional Planning (DITATA) of the Directorate General *Cipta Karya*, which had been responsible for transmigration

---

targets, and the data collected from sending provinces is not a good indicator of the number of people who actually move."

[14]Personal communication with the author, January 1988.

site selection and physical planning, became the Directorate of Program Development (*Bina Program*) in the newly formed Directorate General of Settlement Preparation (PANKIM). The Directorate of Land Preparation was moved to that directorate general, from the Directorate General of Highways in the Department of Public Works. The former Directorate General of Transmigration became the Directorate General for Mobilization and Development (RAHBIN). The budgets of some agencies, such as health and education, were also incorporated into the Department of Transmigration, while other agencies such as Agrarian Affairs and the directorates in agriculture, retained control over their own funds. Under the new department, the staff and functions of the Inspector General and the Secretary General were expanded to improve financial and administrative control (see Fig. 2.1).

There is no doubt that the fourth plan reflected the increasing role of the transmigration program within the national development policy. The target number to be resettled was increased to 750,000 households. The claimed successful outcomes of the third plan had apparently provided a strong argument for the government to increase the target to this level. In the fourth plan, foreign financial support to the transmigration program, particularly from the World Bank, also increased dramatically. However, the appointment of Martono as the minister suggested that the technocrats were losing control of the program. Furthermore, instead of demographic and regional development arguments, national defense and security were highlighted as the important goals of transmigration (Department Penerangan RI 1983, 371–372).

The important role of transmigration for national defense and security was emphasized by the Chief of Armed Forces, General Benny Moerdani, when he delivered a keynote address at the seminar in the National Defense Institute (*Lemhanas*). According to Moerdani, transmigration policy was the only policy within the economic development framework, which had a direct linkage with national security and defense. He argued that it was necessary for the military to be involved in site selection, primarily because transmigration location had a strong relation with the concept of territorial management (*Kompas*, March 8, 1985). An obvious attempt from the military to influence the transmigration policy, among others, was shown by the appointment of Major General C. I. Santoso, a former regional commander in Irian Jaya (1978–1982), as the general secretary of the Department of Transmigration in 1983.[15] In the plan documents, transmigration was discussed in a fairly substantial separate chapter of 58 pages. According to Sjahrir (1987, 193), economist from the Centre for Policy Studies (CPS), the fourth plan showed considerable technical improvement over previous plans, specifically in the utilization of more sophisticated econometrics.[16] However, there were

[15] According to Anderson (1985, 140), a long-standing scholar of Indonesian politics, this appointment should be read as General Moerdani's move to take full control of the "explosive mass-transmigration program in West-Irian."

[16] The Centre for Policy Studies is a nongovernment research institute. Among the senior staff is Professor Sumitro Djojohadikusumo, the "guru" of the New Order's economist–technocrat group.

inconsistencies in the numerical targets between sectors and subsectors. Sjahrir has argued (1987, 193) that these inconsistencies may have been due to the strong influence of the noneconomic considerations in the plan's formulation.

The hopes that had emerged at the beginning of the fourth plan, however, were soon confronted with some hard realities and problems. These problems can be broadly categorized as internal and external problems. The internal problems were related to departmental management, particularly the problem of the unspent budget that reflected the inefficiency of plan implementation. This problem was exacerbated by the gloomy prospects for world oil prices. These internal problems were first felt, according to the minister, around the second year of the fourth plan. The external problems were mostly related to criticisms from abroad, particularly on issues concerning the destruction of tropical forests and unfair treatment of the indigenous populations whose lands were allocated for transmigration. Approaching the end of 1984, the Minister of Transmigration asked Sumitro Djojohadikusumo to act as an advisor to the minister on matters related to program planning and internal management.[17] Among the key persons working for Sumitro was Harun Zain, the former Minister for Manpower and Transmigration. One of the advice by the former minister is related to the problem of unspent budgets. A series of meetings and interviews was conducted with several ministers who were directly linked with the transmigration program, such as Department of Public Works, Home Affairs, Population and Environment, Forestry, and Food Production; the Governor of West Kalimantan; and several Director General. The report was completed in April 1985 and submitted to the minister. It was considered "a reconnaissance phase" and mainly focused on the internal management of the transmigration program (Redecon 1986). The "technocratic" nature of this report and its avoidance of more controversial issues, such as the influence of the strategic military interests on the policy, can be seen as major limitations to the value of the report as a basis for program improvement.

The external problems of transmigration implementation soared rapidly, especially after the government decided to shift the geographic focus of the transmigration receiving areas to the eastern part of Indonesia, particularly to the border between Irian Jaya and Papua New Guinea (PNG). Such a decision, for example, resulted in many technical, as well as social and political problems (Manning and Rumbiak 1989). Due to such issues as environmental degradation and the displacement of the indigenous population, international criticisms began to be heard against the transmigration policy in Indonesia. The World Bank, which for almost 15 years had supported transmigration policy, also became the target of international criticism.[18] In January 1985, a research team from Gadjah Mada University headed by Professor Mubyarto was asked by Minister Martono to observe and report on

[17] A source close to Professor Sumitro told the author (January 1988) that the request to employ Sumitro as an advisor on transmigration had come from President Suharto.

[18] For example, The Ecologist (No. 2/3, 1986) in collaboration with Survival International and *Tapol* published a special issue, which strongly criticized the transmigration program in Indonesia.

transmigration in Irian Jaya. Their observations were presented at a seminar held at the Centre for Rural and Regional Studies, one of the research centers at Gadjah Mada University. Among the issues raised during the seminar and subsequently highlighted by the press was the need to revise and reschedule the transmigration target for Irian Jaya (*Kompas,* February 8, 1985). In addition to the potential social and political conflicts, the original transmigration target in Irian Jaya was economically unrealistic. However, in an apparent response to this criticism, after immediate consultation with the president, Minister Martono strongly emphasized that "*The transmigration target for Irian Jaya will not be reduced but will even be increased.*"[19] Some observers felt that "*Transmigration has become such a fetish with the president that cabinet members dare not try to persuade him to approve drastic changes to it. The premises and goals are non-negotiable*" (Osborne 1985, 25). However, the program was drastically hit as oil prices collapsed in 1985–1986. As a result, the government subsequently reduced the budget for transmigration by 44% in 1986/1987 and a further 65% in the 1987/1988 fiscal years (Tempo, January 16, 1988). With the budget cut, the transmigration target seemed likely to be reduced considerably. This drastic reduction was described by Mubyarto, an economics professor from Gadjah Mada University, as "the will of God."[20]

The substantial cuts in the government budget for the transmigration program had significantly reduced the government's capacity to continue the program. This was mainly because the cost per transmigrant family in the recent period had become remarkably high. In 1983–1984, for example, the cost per transmigrant family was nearly US $6500, whereas at the beginning of the first plan (1969–1970) the cost per transmigrant family was only US $ 577 (Arndt 1983, 66). As extensively reported in the press, many transmigration projects had to be abandoned since the budget cuts. On November 24, 1987, the Minister of Transmigration announced that 16,000 houses built for transmigration had been built in vain because the government had no money to send people to occupy them (*Kompas,* November 25, 1987).[21] This setback attracted many comments and criticisms. Some critics suggested that the transmigration program should be overhauled or the department should be downgraded to its original status as a directorate general.[22] However, rumors surrounding the future of transmigration were dispelled when, in mid-March 1988, the president appointed Major General Soegiarto, a former Head of Social and Political Affairs in the Defense Department, as the Minister of Transmigration.[23] This appointment was further evidence that the economist–technocrat group had

---

[19] *Kompas,* February 26, 1985.

[20] This description was made by Mubyarto at the "Conference on Regional Economy in Indonesia" at ANU, Canberra, February 1987, at which the author was one of the participants.

[21] An unofficial source informed the author that the actual number was over 40,000 houses.

[22] See *Tempo*, December 5, 1987 ("Interview with Minister for Transmigration, Martono"), also, Far Eastern Economic Review, October 29, 1987 ("Indonesia: Resettlement rethink, the government overhauls its transmigration scheme").

[23] Soegiarto, according to Anderson (1985, 142), was one of the "stars" among the A.M.N. (Military Academy) Class of 1960. Recognition of Soegiarto as one among the few "brilliant" generals was

become less influential in the program, but it was also an indication that transmigration would retain an important role in the national scene for many years to come.

The Fifth and Sixth Plans, 1989–1998

The transmigration policy in the fifth plan indicated a new orientation toward more spontaneous transmigration schemes, as well as toward promotion of cash crop transmigration settlements. The fifth plan set the target of 550,000 families to be achieved within 5 years, of which only 180,000 families would be fully supported by the government. The remaining 370,000 families were expected to be assisted through various schemes under the spontaneous transmigration program. The fifth plan also aimed at rehabilitating the poor conditions of existing transmigration settlements. Implementation of the fifth plan proved to be difficult. Besides the fact that the budg*et* allocated for transmigration was obviously limited, another problem also arose as the public image of transmigration had been severely affected by the mismanagement in previous periods. The data on interprovincial migration revealed by the 1990 population census clearly indicated that many migrants were moved to urban rather than rural areas (Mantra 1992). It also reflected a remarkable shift from sponsored to voluntary migration in the mid-1980s as the new economic growth centers developed in several urban areas on the Outer Islands (Tirtosudarmo 1994).[24] In this period, migrants were also attracted by labor opportunities in neighboring countries, particularly Malaysia (Tirtosudarmo 1996). The new minister, as reported in the press, frankly, acknowledged the difficulties faced by his government in recruiting the potential transmigrants. The minister was also aware that people were less willing to move to transmigration settlements, and preferred to move to other places, particularly to urban areas. At the end of the fifth plan, it was obvious that even though the target had already been set exceptionally low, its achievement was generally even lower.

In 1993, with the commencement of the sixth plan (1993–1998), in a clear move to boost the involvement of the private sector in the transmigration program, the president appointed Siswono Yudohusodo, a successful businessman, to be Minister of Transmigration. Although observers generally regarded transmigration as having

---

acknowledged by Goenawan Mohamad, editor in chief of *Tempo* magazine, in a personal communication to the author, March 23, 1988.

[24]The increasing numbers of voluntary migrants to the Outer Islands in the last 10 years have apparently resulted in the increase in conflict between migrants and local people. In 1995, mass open conflicts broke out between migrants and local people in East Timor, Irian Jaya, and Flores, all in Eastern Indonesia provinces (Tirtosudarmo 1995). The latest incident occurred in Sanggau Ledo, West Kalimantan, in the first week of January 1997, between the Madurese migrants and the indigenous Dayak people (*Kompas*, January 3, 1997). Many observers have argued that the economic inequality between migrants and local people is the underlying factor of these ethnic group conflicts (see Tirtosudarmo 1997).

lost its rational justification, President Suharto decided to expand the scope of transmigration to include the resettlement of the so-called forest squatters. The department was then renamed the Department of Transmigration and Resettlement of Forest Squatters. The beginning of the sixth plan also marked the president's statement about various groups of people who were living below the poverty line. The state's rhetoric on raising population groups officially considered as below the poverty line conveniently supported the new task of transmigration (Tirtosudarmo and Handayani 1993). Forest squatters, including many isolated tribal groups in the Outer Islands, had been officially recognized as among those living below the poverty line and become a new target of the transmigration policy. To attract private and business sectors to invest capital in transmigration areas, an economic and business information center was created in 1994. The main idea of this new transmigration policy was in fact similar to one initiated by the World Bank in 1976 in which transmigration was expected to be integrated into the cash crop market economy. The major problem with this venture, however, was the lack of legal and business capacities within the Department of Transmigration. The new minister had indeed pushed the transmigration policy toward a more business and market orientation yet, as the existing system in the Ministry of Transmigration became overly bureaucratized, the ministerial effort to attract the private sector into the transmigration program turned out to be unsuccessful.[25]

Summary

Since the beginning of the New Order, transmigration has become the target for various interest groups, particularly the military, which has perceived transmigration as a vehicle for strategic goals such as national defense and security, and the economist–technocrats who have considered transmigration as a tool to promote the development of remote areas and to alleviate poverty in Java. However, the principal actor had been President Suharto himself. It was Suharto who in 1966 saw the program as an important policy to solve the population problems of Java. At the beginning of the New Order development policy, population distribution was regarded by the economist–technocrats, such as Widjojo, as a low priority. The fact that the president himself gave personal attention to the matter made it impossible for rational technocrats to remove the program from the national development policy. It therefore became an obligation for the economist–technocrats to develop arguments to legitimate the program. They found this in the notion of regional development. From the first through the fourth five-year development plans, transmigration was described as a government policy to resolve the problem of population distribution and to enhance the regional development of the Outer Islands. The

[25] Some observers have even criticized the new attempt as a setback, as it reflects the revival of the old idea of the colonial state's economic strategy.

strategic goal of transmigration for national integration and security, hardly surprisingly, was not always fully articulated.

Transmigration policy was revitalized when the revenue rapidly increased because of the oil boom and further flourished as foreign donor agencies, particularly the World Bank, became deeply involved in assisting the program. Plan documents show that the program has become an important strategy for the achievement of a wide range of development goals. However, the declaration of explicit targets in plan documents primarily reflects bureaucratic procedures in relation to the projected availability of resources and budgets. Numerical targets were not based on any specific economic or demographic considerations. The elusive and multifaceted goals of the policy and the constant changes in implementing agencies were strong indications that the program had never been carefully monitored and evaluated. Furthermore, the unavoidable rivalry and conflicts between the economist–technocrats who viewed the program mainly as an element in regional development, and the political bureaucrats, such as Martono, who were close to the military with its strategic interests, have undoubtedly distorted the plan formulation and the implementation. In this situation, the World Bank, which in the beginning saw transmigration to promote commercial tree crop cultivation, could not avoid the pressure to use transmigration policy as a primary tool for distributing people from Java and Bali to the other islands. The crisis in transmigration since the middle of the fourth plan, although mainly a direct result of the drastic budget cutback, has also been a consequence of overly ambitious target setting and the heavily political element in the program's formulation. The demographic impact of transmigration particularly in reducing the growth of population in Java was almost negligible, except in the third plan when the budget for the program was abundant. At the same time, transmigration also had significant demographic impact on the receiving areas, particularly in Sumatra and to a lesser extent Kalimantan and Sulawesi. The government intention to shift the geographic focus of transmigration policy to the eastern islands, to Irian Jaya, has largely failed as the transmigration policy has been confronted with severe financial difficulties since 1985–1986.

In the New Order period that began after the military achieved supremacy over the communists in the mid-1960s, transmigration began to be redesigned as the part of an economic development strategy. Yet, to perceive transmigration policy as if it were solely a demographic or an economic development endeavor would be naive and misleading. The elusive and multifaceted goals of the policy and the constant changes in implementing agencies were strong indications that the program had never been carefully monitored and evaluated. Furthermore, the unavoidable rivalry and conflicts between the economist–technocrats who viewed the program mainly as an element of regional development and the political bureaucrats who were close to the military with its strategic interests have undoubtedly distorted the process of plan formulation and implementation. In this situation, the World Bank, which in the beginning saw transmigration as a mean to promote commercial tree crop cultivation, could not escape the pressure to use transmigration policy as a primary tool for population distribution. After the mid-1980s, the state's attempt to reorient transmigration policy toward a more market-oriented economy apparently failed as the legal and management capacities were basically lacking within the Ministry of Transmigration. As we have learned from history, transmigration is a policy that elegantly provides the ruling elite with a disguised tool for defending their political and economic interests. In the late 1990s, it was strongly indicated that business and the private sector, particularly the logging and mining companies, were among the groups that would benefit from the new aim of the transmigration program—the relocation of people currently relying for their survival on the forest. The relocation of these people, who are still dominant in the upland regions in Sumatra, Kalimantan, Sulawesi, and Irian Jaya, has had a catastrophic impact on their social and cultural lives. The relocation of tribal groups, which are perceived as potentially undermining the state's political hegemony, for example, the Dayaks in Kalimantan and several tribal groups in Irian Jaya, accommodates the strategic goals of the state to subjugate these latent rebellious indigenous groups.[26] Looking to the future, given its significant instrumental value to the ruling elite, the transmigration program will very likely be retained in the state's policy agenda.

[26]In 1995 and 1996, several riots occurred where indigenous people protested against the exploitation of their customary lands by mining under Freeport Inc., in Timika, Irian Jaya, and by logging in East and West Kalimantan. These incidents, particularly in Timika, have strongly influenced the government to design a resettlement program to relocate these tribal groups from their customary lands.

Chapter 3
The Central–Regional Contexts of Transmigration: Riau and South Kalimantan

> It remains clear that for many years transmigration has been a most visible manifestation of the Indonesian centre into the landscape of the major islands of the periphery.
> *Babcock (1986, 157).*

Riau, Sumatra

In the 1980s, Riau was a huge province located in the Strait of Malacca, neighboring to Malaysia and Singapore. In terms of land area, it is the sixth largest province in Indonesia (nearly 95,000 km^2). Its size is increased by the expanse of Littoral Sea, which lies between the eastern-central coast of Sumatra and the western coast of Kalimantan. This sea covers some 235,300 km^2 and contains more than 3200 islands. Riau is commonly divided into two regions: mainland Riau (*Riau daratan*), the landmass belonging to Sumatra including its offshore islands, and Island Riau (*Riau kepulauan*), the sea area with its numerous islands. A prominent physical feature of Riau is the predominance of rivers. Most rivers originate from the Bukit Barisan Mountain Range in central Sumatra and flow eastward to the coast. Among the dozens of rivers intersecting Riau, four are most important. They are, from north to south, the Rokan, Siak, Kampar, and Kuantan–Indragiri rivers. In mainland Riau, the population lives mainly along the rivers, whereas in Island Riau the population is concentrated along the coast. The geographic position of Riau, which is located on the boundary of other countries (Singapore, Malaysia, and Vietnam), as well as its archipelagic nature, has caused this province to be perceived as being vulnerable to outside influences. In this context, transmigration settlements, particularly in the Natuna Islands, were intended by the central government to form "a security belt" in strengthening national defense. From the strategic and military point of view, Riau Province has therefore become an important region.

In terms of population composition, Riau is a mosaic of ethnic groups. This is somewhat unusual among Indonesian provinces, where a particular cultural identity tends to be strongly associated with each region. For Riau, no single cultural identification is possible, except for a loose application of the term Melayu (Kato 1981). Throughout its history, Riau has been exposed to successive waves of

R. Tirtosudarmo, *From Colonization to Nation-State*,
https://doi.org/10.1007/978-981-16-6437-3_3

population movement. Being situated in the center of Sumatra, along the Straits of Malacca, the area has long been a meeting place for different cultures and relatively receptive to outside influences. The most important distinct migrant ethnic group in Riau is the Minangkabau, who have migrated from West Sumatra over several centuries. Early Minangkabau migrants settled mainly on the eastern flank of the Bukit Barisan Mountain Range (Naim 1979). Recent Minangkabau migrants mostly reside in urban areas. More recent migrants from North Sumatra include Christian Bataks and Moslem Tapanuli Mandailings. They have also generally settled in urban centers. Banjarese from South Kalimantan and Buginese from South Sulawesi have been in contact with island Riau for centuries, but it was mainly after the late nineteenth century that they began to establish communities of their own and to put down roots in mainland Riau. Pioneers in this development were the Banjarese, who settled in the present regency of Indragiri Hilir. According to Hamidy (1981, 2–6), about 45% of the population in this regency are Banjarese. In some districts, such as Tembilahan and Kuala Indragiri, the Banjarese accounts for almost 70% of the population. Bugis migrants have also established some villages in Indragiri Hilir, and there are Bugis seafarers living in Island Riau.

Riau has the largest number of the so-called isolated ethnic groups in Sumatra. These include the Sakai, Talang Mamak, Bonai, Akit, Hutan, and Laut. According to government estimates, in 1985 their total number was 32,058 persons and distributed in 7140 households (Kantor Statistik Riau 1986). Another ethnic group is the Chinese, who are concentrated in cities and large towns of both mainland and island Riau. However, it is not unusual in island Riau for Chinese to live in rural villages, where they sometimes form most of the inhabitants. The Javanese who now live in Riau arrived in several waves. Many came to work as laborers on European estates during the Dutch period. Later, during the Japanese occupation, many were brought over as *romusha* or forced laborers to work on Japanese military construction projects. Recent Javanese settlers have come to Riau on government-sponsored transmigration programs as transmigrants. Despite the diversity of ethnic groups who live in Riau, if there are any cultural elements that give some coherence to the society, they are religion and, to a lesser extent, language (Kato 1981). Except for most of the Chinese, Christian Bataks, and isolated ethnic groups, almost all the inhabitants of Riau are Moslems.[1] As for language, the Melayu Riau, from which the national language, Bahasa Indonesia, is derived, is the lingua franca of Riau.

Riau was formally designated as a transmigration receiving area in 1975 by Presidential Decree No. 29. However, the first group of transmigrants was resettled in 1962 in Siabu, Bangkinang District, Kampar Regency. This group consisted of 58 families (275 persons). A second group of 42 families (212 persons) arrived in 1963 in the same area. According to a study by Hamidy and Ahmad (1984, 10), the government resettled transmigrants in Riau during that period for security reasons,

[1] In 1991, the distribution of people by religion in Riau is as follows: Muslim (87.30%), Protestant (2.30%), Catholic (0.50%), Hindu (1.31%), Buddhist (7.30%), and others (2.70%). Sources: Frederick and Worden (1993, 353).

due to political tensions between Indonesia and Malaysia. Transmigrants were sent to the border regions that were considered by the central government to be underpopulated, to bolster territorial defense in the "confrontation" against Malaysia. However, Hamidy and Ahmad (1984) reported that the development of those settlements was unsatisfactory, partly due to poor planning and hasty implementation. At the beginning of the New Order period, Riau played only a small role as a receiving area for transmigration. The first group of transmigrants in that period, consisting of 98 families (459 persons), were resettled in 1968 at Simandolak in Kuantan–Tengah District, Indragiri Hulu Regency. In 1969, 200 families (866 persons) were sent to the same destination. However, the transmigration settlement in Simandolak was also considered unsuccessful. Hamidy and Ahmad (1984, 11) noted that, just 2 years after they were resettled, some of the transmigrants had to sell their belongings to local people to obtain money to buy food. Some were even described as paupers. For the local population, the hardship of the transmigrants' life in Simandolak provided a bad image of the Javanese transmigrants in Riau. A journalist reported that in the first year of the third five-year development plan (1974/1975–1978/1979), about half of the transmigrants in this area left the settlement as they could not survive in such dreadful circumstances (*Kompas*, March 25, 1980).

The central government began to pay more attention to the province of Riau as a potential area for transmigration in 1971. This was partly because the central government decided to open tidal swamps for a transmigration project to cultivate rice, due to the need to increase rice production. During the first plan, the government reclaimed 12,830 hectares of tidal swampland in South and Central Kalimantan, 2730 hectares in West Kalimantan, 9830 hectares in South Sumatra, 6800 hectares in Jambi, and 900 hectares in Riau (Collier 1979). This undertaking was one of the world's major experiments in marginal land utilization and a classic example of a resource creation strategy (Hanson and Koesoebiono 1977). In Riau, the first transmigration settlement in the tidal swamp area was in Teluk Kiambang, Tempuling District, in the regency of Indragiri Hilir, where 150 families (735 persons) were resettled, followed by another 150 families (732 persons) in 1973/1974 (Table 3.1). In 1974, President Suharto instructed the Department of Public Works to open another one million hectares of tidal swamp. A substantial portion of this was to be used for transmigration settlements. The transmigration program was one of several methods being promoted to expand the agricultural base in the Outer Islands to increase food production. As stated in the first five-year plan document, increasing food production was one of the major targets in the first plan. The government's decision to use tidal swamp was partly because the best lands in the upland areas were already under cultivation by local residents. In many areas, customary (*adat*) law gave land management rights to contiguously spaced local units known as *marga* or clans. In many areas where these rights had been exercised, particularly by shifting cultivators, the problem of alienating sufficient land for transmigration was more serious. Thus, in general, the upland areas of Sumatra presented difficult land rights situations for new settlements.

Since the beginning of the second plan, regional development and the establishment of growth poles outside Java have become the major preoccupation of

Table 3.1 Transmigration in Riau (1962–1987)

Year/arrival	Location	No. of families	Year/transfer
1962–1963	Siabu	100	1976
1968–1969	Simandolak	298	1977
1971–1975	Tidal areas	500	1979
1977–1979	Tidal areas	900	1982–1985
	Coastal areas	2000	1987
1979–1981	Upland regions	19,417	
1981–1983	Upland regions	15,772	
1983–1987	Upland regions	17,509	

Source: Kanwil Transmigrasi Riau (1987)

Notes: The figure for transmigrant families in this table should be used with caution because it was taken at the beginning of the resettlement. In the course of time, some transmigrants probably left and new migrants arrived and resettled

In January 1988 (at the time of fieldwork), settlements that opened after 1978 were mostly still under the responsibility of Department of Transmigration

Indonesian planners. Regional development has become the fundamental justification for the transmigration program. Hanson (1981) considered the tidal swamp development effort to be supportive of both agricultural sectoral goals and regional development goals. Another element in regional development was the improvement of security and avoidance of disturbances in sensitive and conflict areas. According to Hanson (1981, 224), the central government was anxious to settle transmigrants in such areas to reduce opportunities for illegal activities, such as smuggling. In coastal Riau, particularly at isolated outposts and along the network of swampland channels or rivers, there were ample opportunities for many kinds of smuggling. These considerations have been factors in the emphasis on coastal development. During the third plan period, 3100 families (12,395 persons) were resettled in Riau, most of them in the coastal areas (Table 3.1).

The transmigration target for Riau was increased significantly with the beginning of the third plan. The increase was due to the rapid increase in the target at the national level and the judgment that Riau has the potential for resettling transmigrants. In the third plan, the national target was 500,000 families, of which Riau was to receive 50,000. However, the planned expansion of the transmigration target confronted an immediate problem: the availability of suitable land for resettlement. By the end of the third plan, it was evident that resettlement in tidal swamps had created many problems. As a result, it was recommended that this type of project should be de-emphasized in the future. Since then, the central government has decided to shift transmigration settlement into nonirrigated areas in the upland regions (Table 3.1). Since fertile land in the upper regions mostly has been occupied by the local population, the new settlements for transmigration were resettled in the areas with poor soil conditions. In addition to food crops, since the beginning of the third development plan the central government introduced the tree crop type of transmigration. In Riau, an oil palm plantation was established in Tandun and Sei Tapung in the regency of Kampar. It was followed by a rubber tree plantation in Air

Molek in the regency of Indragiri Hulu. The province of Riau was given special attention by the central government partly because of national security as Riau has an open border with other countries in the north, including Malaysia, Singapore, and Vietnam. In 1981, the government moved 350 families (1627 persons) to the Natuna Islands, and in 1982 another 600 families (2425 persons). The government explicitly stated that the aim of this resettlement was to provide "a security belt" against foreign intruders. One researcher from the Bogor Institute of Agriculture (IPB), who had conducted field research in transmigration settlements in Natuna, gained a strong impression that living conditions among these transmigrants were extremely poor indeed, particularly due to the exceptionally low level of agricultural productivity. He considered the government's expectation that the transmigrants could act as a security belt when they had to live under such conditions that were merely wishful thinking.[2]

South Kalimantan

South Kalimantan, on the other hand, is one of the smaller provinces (37,000 km^2) and by far the smallest of the four provinces into which Kalimantan, the largest island in Indonesia, is now divided. Its topography is almost entirely flat, with only a small area of mountainous and hilly country in the eastern part of the province. The flatness is such that the Barito River, the largest river, is navigable 100 km upstream, rising only 0.5 m over that distance. The Barito with its tributaries connects the south and north of the province. Banjarmasin, the capital, is at sea level, and the high tide regularly floods part of the city and the surrounding country. More than half of the area of one million hectares under swamps and lakes is influenced by tidewater. South Kalimantan may conveniently be divided into three regions: the western lowlands; a central region of foothills and mountain country; and the eastern lowlands. The western lowlands comprise the lower reaches of the Barito River and its southern tributary, the Martapura, and the floodplain of another tributary, the Negara. Bordering the western foothills of the Meratus Mountains is a strip of alluvial soil, which comprises the most fertile area for rice production and is thus the most heavily populated rural district. It contains the important towns of Barabai and Kandangan, in the regency of Hulu Sungai Tengah and Hulu Sungai Selatan, and Rantau in the regency of Tapin. The western foothill region has been exploited, mainly by shifting cultivation methods, for over a century. Parts of the foothills have long been under smallholder rubber in the northern regency of Tabalong and Hulu Sungai Utara. In the most southern regency of Tanah Laut, where grassland is most extensive, much scope exists for livestock development. Along the southern and eastern edges of the mountains, human occupation has been more recent. The eastern

[2]Personal communication, January 1988.

lowlands, occupied by one regency, Kotabaru, constitute South Kalimantan's "frontier zone."

The dominant ethnic group in South Kalimantan is the Banjarese. South Kalimantan is sometimes referred to as the Banjarese heartland. The Banjarese is in fact a mixture of the indigenous population (the Dayaks), refugee Malays, who migrated after the downfall of Melaka, and the Javanese, who fled Demak after the attack of Sultan Agung of Mataram (Saleh 1962, 17). The manifestation of such a mixture is also seen in the Banjarese language, which is a patois of Malay, Javanese, and indigenous languages. The penetration of Javanese into the Banjarese culture can also be seen from the Banjarese aristocratic status and associated language patterns, organizational arrangements, food, dances, and costumes (Saleh 1962, 62). The Banjarese, like the Minangkabau and the Bugis, has a high propensity to migrate. Traditionally, they have migrated to Sumatra, Malaysia, Java, and Maluku. In the 1930 census, about 13% of the Banjarese were counted outside South Kalimantan, most of these in Sumatra (9.6%) and the Malay Peninsula (2.5%) (Basri 1988). Naim (1979), based on the 1930 census and his own survey, argued that, in contrast with the Minangkabaus, the volume of Banjarese out-migration had been gradually declining.

Basri (1988) argued that Banjarese migrants are economically conservative because the majority of them continue to work in agriculture, as they did in their homeland. The majority of Banjarese migrants in North Sumatra, Riau, Jambi, and Malaysia are rice farmers and rubber or coconut plantation owners, with only few working as laborers. Many Banjarese who migrated to Sumatra originated from the regencies of Hulu Sungai, Kandangan, Barabai, and Tanjung (Basri 1988, 48). The first waves of Banjarese who migrated from South Kalimantan were motivated by political considerations after their defeat by the Dutch. However, their success in opening new lands attracted more movement and formed chains of migration that persist today. The Dayaks are concentrated in the regencies of Hulu Sungai Utara, Tabalong, and Barito Kuala, whereas the Bugis migrants from South Sulawesi have formed communities in the regency of Kotabaru. The Javanese and Madurese, who mostly migrated under government assistance projects, are settled in a variety of small transmigration areas, particularly in the regencies of Barito Kuala, Tanah Laut, and Kotabaru. Isolated ethnic groups are scattered in remote areas, such as in the regencies of Tanah Laut, Tapin, Hulu Sungai Selatan, and Tabalong. Most of them still practice shifting cultivation (*ladang berpindah*), as well as hunting and gathering. The government estimated their number in 1983 at almost 30,000 persons. Apart from the isolated ethnic groups and the Chinese, the majority of the population in South Kalimantan is Muslim.[3] According to Saleh (1962), Islam was brought in and spread through traders and Islamic leaders from Demak on the north shore of Central Java.

Whereas Riau has become a transmigration receiving area only recently, South Kalimantan has a longer history of population resettlement. It was in the early 1930s

---

[3]Muslim comprises 97.61% (Frederick and Worden 1993, 353).

that the Dutch government decided to extend their colonization schemes from Sumatra to Kalimantan.[4] In Kalimantan, the Dutch government experimented with two kinds of landscape: tidal swamplands and the alang-alang-covered uplands. For the tidal swamplands, they selected an area to the south of Banjarmasin, which later called as Purwosari, and for the alang-alang-covered uplands, an area in the northeast of Banjarmasin was chosen, which later called as Madurejo. According to Pelzer (1945a, b, 223–225), in July 1938 the first group of 95 families was placed near the village of Kertak Hanyar to cultivate 100 *bouws* of swamp rice field that had been leased from the local population. The experiment was successful, and the Dutch government felt sufficiently encouraged to send a second group of 106 families in 1939. This group came early enough to help with the harvest, and the settlers of 1938 were then transferred across the Barito River and settled on their own land south of the Serapat Canal and southwest of Banjarmasin. The colony, given the name of Purwosari, comprised an area of 34,000 hectares. The second group was brought to a local village on the banks of the Lauk River to learn the new technique before they joined the settlers of Purwosari. In May 1940, the third year, a third group of 198 families was added to the Purwosari project. The planners of this colony estimated that 15% of the land would be occupied by villages, canals, and roads, leaving 28,900 hectares of swamplands for the cultivation of rice. The other scheme, Madurejo, also differed fundamentally from the colonies in Sumatra. In Madurejo, dryland was cultivated exclusively, and Madurese villagers were chosen to be the settlers. The colony, located near Pengaron in the Meratus Mountains, included 16,000 hectares of cultivable land, none of which could be irrigated. This was no drawback in the eyes of the Madurese settlers, since the peasants of Madura surpass all other Indonesians in the tilling of *tegalan* (dryland). The Madurese are excellent stock breeders. The goal in Madurejo was to establish a system of mixed farming. The colony was started in 1938 with 100 families, and these were joined in 1939 by 194 additional families (Table 3.2).

The population settlements in Purwosari (1937) and Madurejo (1938) are regarded by some observers as an example of successful resettlement. Collier (1980 cited in Potter and Hasymi 1987, 16) strongly argued that the success of colonization in Purwosari was because the Javanese followed the type of cropping system devised for the tidal swamps by spontaneous Banjarese settlers, who had earlier opened up sections of the swamps for the production of rice and tree crops in a system which he described as productive and self-sustaining. However, an interview by the author in February 1988 with some former colonists in Purwosari gave the strong impression that the tight control by the Dutch over the lives and agricultural practices of the settlers was perhaps the primary factor in making the settlements sustainable. For example, for 2 years after their arrival in the area, the Dutch officers did not permit the settlers to move freely from the settlement areas. In addition, the

[4]Sjamsu (1960, 59) noted that, in 1920, the Dutch sent 250 families from Central Java to a location near Barabai. After 2 years, this settlement was abandoned as a large number of colonists left the area or died from malaria.

Table 3.2 Transmigration in South Kalimantan (1938–1988)

Year/arrival	Location	No. of families	Year/transfer
1938–1940	Purwosari	399	1960s
	Madurejo	410	1960s
1953	Takisung	724	1960s
1957–1967	Upland regions	250	1970s
1969–1974	Tidal land areas	1859	1980s
1975–1978	Upland regions	6631	1980s
1979–1981	Upland regions	11,096	
1984–1988	Upland regions	5332	

Sources: Pelzer (1945a, b, 224–264); Bappeda Kalimantan Selatan (1986, 2); Djohansyah (1986, 11–13); Kanwil Transmigrasi Kalimantan Selatan (1987, 15)

Notes

1. The figure for the number of the transmigrants in this table should be treated with caution since the figure was taken at the beginning of the resettlement period. The actual figures could be different as some transmigrants probably had abandoned their settlements or some spontaneous migrants resettled in those areas

2. As also happened in Riau, by the time the author was conducting fieldwork (February 1988) the majority of settlements, which were opened after 1978, were still in the hands of the Department of Transmigration

agricultural activities that could be practiced by the settlers, such as the crops to be grown, were also strictly selected and defined by the Dutch. One ex-colonist said, "It was like living in prison."[5] To judge the success of the colonization policy in terms of the current situation is also misleading for two reasons. First, the original settlers still living are the remnant of those colonists who survived after almost 50 years of hardship, and whose human and social costs are impossible to calculate. Second, the settlements have attracted considerable numbers of new migrants, many of whom have brought capital from their place of origin.

The arrival of the Japanese in the Indonesian archipelago in 1942 put an end to all colonization projects. Most of the settlements already established suffered considerably during the 3 years of occupation; in many cases, farmers were taken from colonization villages to provide labor for Japanese war projects (*romusha*). After independence, the government of Indonesia resumed the program of colonization. The first group of transmigrants was sent to South Kalimantan in 1953 to Takisung District, Tanah Laut Regency. They comprised 724 families (2595 persons) (Bappeda Kalimantan Selatan 1986, 2). After Takisung, the government began to open tidal irrigation projects (*proyek pasang surut*), the first at Tamban in 1957,

[5]Lesley Potter of Adelaide University who was conducting research on population mobility in South Kalimantan, found, among other things, a Javanese settlement at Guntung Ujung, Kecamatan (District) Gambut, about 12 km south of the main Banjarmasin–Martapura road. According to Dr. Potter, there were about 40 households, some of them descendants of a group who fled from the colonization at Purwosari during the Japanese occupation of the 1940s. They left the settlement because they found life in Purwosari too difficult (a personal communication with Dr. Potter, December 1989).

followed by two more tidal projects at Marabahan in 1959 and Balandean in 1961. All of these projects are located in the regency of Barito Kuala. Two other projects were established in the regency of Kotabaru: first in Berangas in 1960 and second in Sebelimbingan in 1961. From 1962 until 1967, the government moved only around 250 families to South Kalimantan. A survey of the regional economy (Partadiredja 1970, 51) reported that, by 1968, only two-thirds of transmigrants arriving after independence still resided in the transmigration areas. The others had left for other places to find employment. The report also found that, out of five areas designated as transmigration areas since 1959, only one was officially regarded as developed and four as "backward."

South Kalimantan was formally designated as a destination for transmigration in 1972 by Presidential Decree No. 2 (Department of Manpower and Transmigration 1979). During the first plan, about 8000 persons were resettled in seven transmigration areas. Around half were in Barambai in the regency of Barito Kuala in 1969, and the other half were in 1973 in Tambarangan, Hatungun, Sidodadi, Miawa, Parandakan, and Ayunan Papan in the Tapin Regency. According to a study by a team from Gadjah Mada University, the transmigration settlement in Barambai was relatively more developed than the other settlements (Guinness 1977). In the second plan, the government moved around 28,000 persons into the dryland areas in the regency of Kotabaru. The majority were in Tajau Pecah in 1975 and in Sebamban I in 1978. The others were in Suato, Hatungun, Batu Tungku, Masingai, and Belawan. In Sebamban I, in which around 8000 persons were located, an agroeconomic survey showed that many transmigrants were engaged in nonfarm employment, such as logging and the informal sector (ORSTOM 1984 cited in Widayatun 1988, 16). During the third plan, 77,000 sponsored and around 9000 spontaneous transmigrants were resettled in South Kalimantan. In addition to the transmigration program based on food crops, after 1981 a tree crop transmigration scheme, mainly rubber, began to be introduced in South Kalimantan. These projects were located in Batulicin, Pamukan, Paringin, and Muara Uya. At the beginning of the fourth plan (1984–1985), a transmigration area associated with a sugar plantation was established in Pelaihari, Tanah Laut Regency.

Central Decisions and Regional Implementation

In Riau and South Kalimantan, the transmigration program has been implemented since the first five-year development plan of the New Order period in 1969. However, the momentum of the program increased markedly from the beginning of the third plan period in 1983. At the national level, this marked change was reflected through at least three important developments. The first was the commitment of the World Bank to financially support and assist the transmigration program on a substantial scale. This dates from 1979, when the government and the World Bank agreed to the second bank-assisted project that was to be a model for transmigration in the third plan (World Bank 1988, 161). Second, a total of 500,000 families were targeted to move in the third plan. Third, a Junior Minister for Transmigration was appointed to assist the Minister for Manpower and Transmigration although he reported directly to the president. In addition, a coordinating body (*Bakoptrans*—see Chap. 2) was set up to coordinate the implementing agencies for transmigration.

Based on their study in South Kalimantan and South Sulawesi, Guinness (1977, 137) noted that, although the government had established a coordinating body at the national and regional levels, this body was not able to reduce the tensions and conflicts between the implementing agencies. The only experience of workable coordination, according to this study, was shown during the preparation of the Wonogiri–Sitiung Project in 1976. However, from another study (Gondowarsito 1986), the seemingly better coordination among the implementing agencies in the Wonogiri–Sitiung Project was because President Suharto had given his personal attention to the project, presumably because Wonogiri happened to be in the area of his childhood home. Because of this privileged status, the utmost involvement and cooperation on the part of various government agencies (namely the Departments of Manpower and Transmigration, Public Works, Communications, and Agriculture) were enlisted (Gondowarsito 1986, 89–90). The main reason that eventually led the central government to rapidly increase the scale of the transmigration program was the increase in national revenue from oil, as well as the increasing confidence among the economist–technocrats in their ability to expand the development programs on a nationwide scale. The central government decision to resettle 500,000 families in the third five-year plan compelled the designated provinces to accept a larger number of transmigrants than had ever been experienced before. This obligation in turn created a heavy burden and many problems for the regional governments.

In Riau, prior to the commencement of the third plan, the regional government had planned for 50,000 transmigrant families to be accepted for the next 5 years and had already allocated areas for transmigration settlement. Suddenly, and without any consultations with the regional government, the new Minister for Public Works, Poernomosidi, increased the target to an unprecedented 100,000 families. The central government, through their vertical line agencies in the region, particularly Public Works, began to select settlement areas, clear the land, and construct roads. The Transmigration Regional Office then started to build the houses and social facilities, such as markets and mosques. During the project's development, almost all consultations were carried out directly between the central government's implementing agencies and the governor's office, largely bypassing *Bappeda*, which was supposed to coordinate all development projects in the region.[6]

The bureaucratic elites in South Kalimantan, especially the planners, have realized that partly because of the soil conditions their province could not receive large numbers of transmigrants from Java. According to Sam'ani, the Head of Research Section in the *Bappeda* office, the findings of the third plan evaluation strongly indicated that for the fourth plan South Kalimantan should receive no more than the target in the third plan, which were about 17,000 families. He argued that even to

[6]Interview with the Vice Head of the *Bappeda* of Riau, January 1989.

realize such a number would be a difficult task, requiring considerable effort. It was therefore beyond the regional capacities to accept a bigger target number. For the fourth plan, therefore, *Bappeda* planned for only 17,000 families. Unfortunately, before it was officially publicized, one senior official from the Department of Transmigration in Jakarta detected this plan during his visit to South Kalimantan. He claimed that the figure was too low and could be interpreted as an indication that South Kalimantan did not support the national policy on transmigration. As a result, in the Development Plan book, 50,000 families were listed as the province's target for the fourth plan. However, by February 1988 only about 13.5% of this target was realized.[7]

The government approach to transmigration during the third plan period was known as "Plan as You Proceed" (PAYP). At the time the transmigration program was reorganized in 1978, the responsibility for the selection and evaluation of large sites was assigned to the Directorate of City and Regional Planning (DITATA) within the Department of Public Works.[8] Plan As You Proceed, initially promoted by World Bank consultants, denotes an approach that was adopted especially by the planners in DITATA to accelerate the completion of land clearing.[9] In the PAYP, site investigation and land development proceeded simultaneously. This approach was adopted partly because the government had been forced to achieve an overly ambitious target for the third plan. Under this approach, site selection was made without prior detailed feasibility studies. The sites were selected mainly based on topographic maps of land slopes. Based on those considerations, the contractors were assigned to land clearing and construction of the main roads. It was then the responsibility of the Transmigration Regional Offices to construct houses and other public facilities for the transmigrants.[10] A *Kompas* daily correspondent in Pekanbaru noted that the contractors who won the projects from Kanwil Transmigrasi often ignored the existence of the regency government or its subordinates during project implementation. All jobs were directly negotiated with the provincial-level government. However, in the event of a dispute or conflict with the local population, for example, because of unfair land compensation, the contractors eventually turned to the regency or village-level government for help or advice.[11]

Coordination between DITATA, which was under the Ministry of Public Works, and the Transmigration Regional Offices was not easily achieved. There was a strong tendency for DITATA to finish their job well before the transmigration offices were ready to follow it up. Among the planners in DITATA, such an acceleration of land

---

[7]Interview, February 1988.

[8]DITADA apparently had no previous experience in agricultural planning and implementation.

[9]The information regarding the role of the World Bank as the promoter of PAYP was obtained through a personal communication from Mr. Wong Tai Chee, a consultant with a Malaysian firm in charge of land development for transmigration in Riau, November 1988.

[10]Interview with Abdul Muin Abbas, a former field supervisor during the PAYP phase in South Kalimantan, February 1988.

[11]Interview, December 1987.

development was known as "saving" (*menabung*). It meant that land development was accelerated and continued without consideration for whether Transmigration Regional Offices had been prepared to construct the houses and other facilities (Hamzah 1980, 5). As a result of the delays in the settlement preparation, for example, Hendra Sasmita, the Head of the Transmigration Regional Office in Riau, admitted that only 20% of the target of 9500 families for the first year of the third plan was realized (*Kompas,* March 11, 1980). According to the Deputy Head of *Bappeda,* Riau, because of the PAYP approach, a considerable proportion of the sites proved to be inappropriate for transmigration settlements. The soils in many areas proved to be unsuited for agricultural cultivation due to the abandonment of the detailed soil structure studies. In 1982, a daily newspaper reported that 1836 transmigrant families from 14 settlements in Riau were in a critical condition resulting from flooding of their lands (*Sinar Harapan*, October 7, 1982). According to one staff member of the Mobilization and Development Division (RAHBIN) in the Transmigration Regional Office in Pekanbaru, it was no accident that the majority of the settlers had left the transmigration settlement in the Belila area because of floods. In fact, before the area was opened for transmigration, the Head of the Transmigration Regional Office had already recognized that the area was not feasible for resettlement. However, the Head of *Kanwil* at the time insisted that the project be "coercively" implemented (*dihajar*, literally means "to crush") because there was an order from the center under the so-called crash program to achieve the transmigration target by every means.[12]

Another problem resulted from the PAYP, and target-oriented approach was disputes over the land required for transmigration among the different government sectoral agencies. The land sites that had been claimed for transmigration settlements and were being cleared had often been originally planned for other purposes, for example, as forest conservation areas. In Riau, land clearing for WPP (partial development areas) X in Bangkinang Regency had to be suspended as the Forestry Technical Office (*Dinas Kehutanan*) in Riau claimed that the area was already reserved for conservation. At the end of the dispute, which was handled by *Bappeda,* the claim of the forestry office was approved (*Kompas,* November 30, 1982). In South Kalimantan when the transmigration program was implemented under the PAYP approach, land disputes commonly occurred with the forestry offices. Abdul Muin Abbas explained that, as a field supervisor, he required a tough character because a quick decision and sometimes even strong pressure were needed to win the dispute with other agencies.[13] According to Muin Abbas, because of frequent disputes between agencies, the regional government in conjunction with other agencies, which used land, such as Department of Public Works, Agriculture, and Transmigration, began to construct a kind of land-use map to mark the forest boundary according to its functions. The map was later known as "Map of Forestry Agreement" (*Peta Hutan Kesepakatan*). At the bottom of the map was the signature

[12] Interview, December 1987.

[13] Interview, February 1988.

of the head of each agency involved. However, because the map was prepared only to provide a broad guideline, particularly during the PAYP approach, in practice disputes between agencies still occurred. Part of the reason was that the central government, particularly Department of Public Works, often intervened directly through their sectoral agencies in the province without consulting the *Bappeda* or other sectoral agencies.[14] The problems of land unsuitability for transmigration settlements, which continued during the fourth five-year development plan, were unavoidable because the land that had been allocated for settlements in the third plan was mostly poorly selected and hastily studied during the third plan period, particularly when the PAYP approach was adopted.[15] The result of poor planning and hasty implementation of the transmigration programs in Riau and South Kalimantan is readily apparent from two indicators: first, the extent to which the target was realized and, second, the very low number of villages that had been transferred from the transmigration department to the regional government. In 1988, for example, no transmigration villages that were opened since the beginning of the third plan were transferred to the regional government.

Regional Responses to Central's Plan Formulation

Regional development, which, on paper, is the responsibility of the Regional Development Planning Board (*Bappeda*), has been firmly controlled by each sectoral agency (*Kanwil*). The regional plan for sectoral agencies in fact is largely decided by the national-level planning board in Jakarta (*Bappenas*). If the regional-level agencies have any opportunity to formulate a regional plan, it is necessary that this be in accordance with priorities at the national level. In the words of one planner in South Kalimantan: "*We should know what musical instrument is being played in Jakarta. For example, if Jakarta plays the gamelan, we in the region also have to play the gamelan.*"[16] The bureaucrats in Pekanbaru or Banjarmasin are well aware of the dominant role played by the central government, either by the ministerial offices or by the National Planning Board in Jakarta. This is reflected in their attitudes and responses in attempts to influence the decision-making process regarding development in their regions. Among the factors that they perceive as crucial to development in their regions is the allocation of the development budget. As noted earlier, the central government allocates 70–80% of the Regional Development Budget (APBD)

[14] Interview with Anwar Rahman, *Bappeda* Riau, January 1988.

[15] In the case of South Kalimantan, the Department of Transmigration, funded by the World Bank, had employed a British Consulting Firm to conduct a detailed land survey in South and East Kalimantan. The report of this survey was unfortunately released only in May 1987. The findings from this survey have never been used since many of the transmigration projects were already being implemented from around the early 1980s.

[16] Interview with Ismet Ahmad, the Dean of Agriculture Faculty, University of Lambung Mangkurat (also the Head of Governor's Task Force), February 1988.

in all provinces in Indonesia. Since the allocation of the development budget will be determined according to the priorities of the five-year development plan, it is especially important for each regional government to make sure that as many as possible of their regional development priorities are mentioned in that document.

There are at least two opportunities for the regional government to influence the decision-making process in the central government. The first is at the time when the central government is preparing the draft of the five-year development plan. This event usually commences around the end of the fourth year of the current development plan. The second is around January and February in each year when the budget for the following financial year is being prepared. The approach used by the regional governments in their attempt to influence the plan formulation in the central government is popularly known as "lobbying the center." The Deputy Chairman of *Bappeda* in Riau explained that his agency's major attempt to influence central government decisions was by inviting experts from the Faculty of Economics of the University of Indonesia in Jakarta to conduct an evaluation of the implementation of the fourth plan and at the same time to prepare the details of the fifth plan. The Deputy Chairman (himself an economist who graduated from the University of Indonesia) apparently knows very well that by inviting economists from this university he will be able to bring Riau's economic issues to the inner circle of decision makers in Jakarta, which is dominated by economists from the University of Indonesia, some of whom are ministers who meet regularly at the university to discuss the latest economic development issues. Another reason for inviting economists from Jakarta is because they are more likely to be listened to when speaking about the problem in Riau than are the people from Riau.[17] The issue of revenue from oil, which only brings a small economic return to Riau, is an example of an issue that would be regarded as very sensitive if it was raised by local people. Through the economists from the center, it was hoped that the decision makers in Jakarta would be more willing to accept the fact that Riau needs a larger share of the national revenue.[18]

The bureaucrats in South Kalimantan used a different approach in lobbying the central government to pay more attention to development in their province.[19] According to Marliani Djohansyah, in 1986 the governor had decided to form a Task Force consisting of academics and intellectuals concerned with local problems. Their main task was to advise the governor on development matters. Such a role

[17] The evaluation and planning of the previous *Repelita* usually was carried out by local economists from the University of Riau, Pekanbaru.

[18] From a conversation with one researcher from the University of Riau (December 1987), it was known that the principal researchers who conducted the study on Riau were Dorodjatun Kuntjorojakti, Director of the Institute of Economic and Social Research, Faculty of Economics, University of Indonesia, and Iwan Jaya Azis, a researcher and regional economic expert, who strongly supported the idea of decentralization of regional development in Indonesia.

[19] The information regarding this issue was gathered from separate interviews with Ismet Ahmad and Ali Hasymi from University of Lambung Mangkurat, and Sam'ani from *Bappeda*. They were all the members of the Governor's Task Force.

should be played by the *Bappeda,* but the governor initiated this Task Force because he considered that *Bappeda* was often too slow in responding to his requests. Another duty given to the Task Force was to lobby the central government to direct more attention to development in South Kalimantan.[20] In a manner rather different from what happened in Riau, South Kalimantan preferred to send its messages to Jakarta through contact persons or close friends who held key positions, for example, in *Bappenas,* the Ministry of Finance, or the State Secretariat.[21] In January 1988, when the author was in the field, the central government had begun to prepare the fifth development plan. The lobbyists from South Kalimantan had therefore started visiting Jakarta seeking to ensure that South Kalimantan would be given priority within the forthcoming fifth plan. South Kalimantan was seeking a larger budg*et al*location for development. According to one member of the Task Force, to attain this goal, the payment of "a greasing fee" (*uang pelicin)* to central government officials was a common practice. He said that "*If you want to reach your destination faster, you should use the toll road. It means that you must pay.*"[22]

Marliani Djohansyah, a local university's lecturer, who conducted a study in several transmigration settlements in the regency of Kotabaru, stated that a major problem in transmigration program implementation in South Kalimantan was that the government officers at the regency level could not change the policy during implementation. Everything had been rigidly decided by the higher-level offices in the center (in this case either in Banjarmasin or in Jakarta). The Head of *Bappeda* at the regency level even said that "*I think everybody in Kotabaru knows that the areas which have been chosen for transmigration are not suitable for cultivation. But I could not change anything since it has been decided from above.*"[23] In Sebamban, Batulicin, where transmigrants had been resettled for almost 5 years, the coordination between agencies, such as those responsible for cooperation, health, and transmigration, occurred only in the form of formal meetings, in which agencies informed each other about their plans. Since most of the funds were controlled by the transmigration office, the other agencies could only follow what they were told by the transmigration officials. As described by Marliani Djohansyah, other agencies working in transmigration areas usually followed the lead of the transmigration officials. Any resentment toward or dissatisfaction with the transmigration officials was usually not openly articulated, although on one occasion an open protest was made by some young medical doctors (*Dokter Inpres*). They protested that the funds allocated to purchase medicines had been misused by transmigration officials. As a result of this protest, the official who apparently misused the money was transferred to another location. Another example of very rigid and centralized program

[20] Interview, February 1988.

[21] Interview with Sam'ani on February 1988 provides information that on January 25, 1988, Ali Hasymi and Asnawi, and on January 28, 1988, Ismet Ahmad and Sam'ani went to Jakarta in an effort to lobby the decision makers in Jakarta.

[22] Interview, February 1988.

[23] As told by Marliani Djohansyah to the author during an interview, February 1988.

implementation was the provision of salted fish to the transmigrants. The central government determined that this had to be imported from Surabaya. As a result, the cost of the salted fish was three times higher than if it had been purchased in the local markets, which had an abundant supply.[24]

Any observers who visit transmigration settlements in Sebamban in the district of Kotabaru, which during the third and fourth Plan periods became the focus of transmigration activities in South Kalimantan, will immediately be aware of the clear failure of transmigration implementation. The settlements, although they have been inhabited by migrants from Java since 1980, show truly little progress. In February 1988 (at the time when the author visited the areas), these transmigration villages were still under the Department of Transmigration with no indication when they were likely to be transferred to the provincial government. Since the mid-1980s, the press has continuously reported the plight and difficult lives of the transmigrants in these areas. For example, press reports documented many people who abandoned their settlements, most of whom returned to their places of origin in Java. In March 1986, a press report showed that hundreds of transmigrant families who were resettled in the Sebamban II project since 1980 were in serious difficulties. Their agricultural lands could not produce anything except cassava, and many of them even had to sell the tin from their roofs and wood from the walls of their houses just to buy rice (*Kompas,* April 8, 1986). An interview by the author with an apparently successful transmigrant trader in Sebamban II in February 1988 provided information about the common phenomenon in transmigration areas. According to him, to survive in an environment such as Sebamban, transmigrants must find jobs outside their settlements in off-farm activities, for example, working as laborers in logging activities or taking a risk, with two or three friends working under contract with their "boss" to collect sandalwoods (*kayu gaharu*) in the forest. The problem, he said, was that productive jobs were extremely limited and so many people were looking for work. As a result, the employers or the "boss" have often paid them an exceptionally low wages, but of course there was no alternative for these people other than to accept this low payment. This informant also stated what has become a public knowledge in these areas that some wives of transmigrants were engaging in prostitution, again organized by a "boss" in the nearby town, to earn some money to enable their families to survive.

Unresolved Problem of Policy Coordination

During the New Order period, formally, the principal planners of transmigration policy were the economist–technocrat group, which was institutionally represented in the National Development Planning Board *(Bappenas)* in Jakarta. In formulating the plan, *Bappenas* was responsible for coordinating the several ministries dealing

[24]Interview, February 1988.

with transmigration policy, including Department of Transmigration, Home Affairs, Public Works and Finance, and the Regional Development Planning Boards (*Bappeda*). As argued by Glassburner (1978, 29), although *Bappenas* is the center of activity at the time of writing the plan, this agency is not granted a real central management system, which would allow it to control and operationalize the formulated plan. During the implementation of the plan, changing policy directions without consulting *Bappenas* could easily take place, mainly because the actual power holders are beyond the control of *Bappenas*. In the case of transmigration, the president or the ministers are often more influential in directing the policy implementation. In the policymaking process, the roles of provincial-level government apparatus, as well as the regional-level implementing agencies (*Kanwil Transmigrasi),* were largely confined to the provision of the information needed by the planners and decision makers in the central government. Among the planners and bureaucratic elites in the regions, the dominant position of the central government in the policymaking process was fully recognized. In attempt to influence the decision-making process of the central government, several strategies were adopted by the bureaucratic elites in the regions.

The organization responsible for the implementation of policy is in most instances a unit of the government bureaucracy. Smith (c.1987, unpub.) argued that bureaucracies are not machine-like entities that faithfully and diligently go about their tasks of policy implementation in a neutral, programmed, mechanistic manner. Bureaucratic agencies are controlled and staffed by individuals who have their own feelings, attitudes, and behavior patterns. Organizations, themselves, operate in an environment of overlapping responsibilities with other agencies and have their favored clients and their friends, as well as enemies and adversaries. "Bureaucratic politics" is a concept often used to describe organizational operations within this highly charged bureaucratic context. Three key variables of the implementing organization should be considered in the implementation of the policy. These are the structure and personnel of the organization, the leadership of the administrative organization, and the implementation of program and capacity. The structure and personnel of the transmigration implementation agency have not been very stable since it constantly shifted from one ministry to another for each new five-year development plan. As a result of this change within the government bureaucracy, the structure of the organization was also subject to frequent change. Although the personnel of the transmigration agency were mostly the same people, they experienced continual uncertainty because of the frequent changes in ministry that authorize the agency.

Problems of coordination were experienced, not only between implementation agencies from different departments but also within the Department of Transmigration itself. For the implementation agencies in the region, the problem arose whenever there was a conflict or confusion about central office decisions, for example, between instructions from the minister and the general secretary. This happened in Riau with regard to arrangements for transporting rice for transmigrants. Initially, an instruction letter signed by the minister stated that transportation should be arranged by the central office. Before this instruction could be implemented, the region received another instruction from the general secretary, which said that the regional office was responsible for transport arrangements. Several months passed before the confusion could be resolved. In the meantime, the transmigrants who had been waiting for their rice ration suffered.[25] According to an agricultural engineer who worked as a staff member of the Program Development Division (*Bina Program*) in the Transmigration Regional Office in Pekanbaru, the accuracy of transmigration statistics in many government reports was questionable. This was partly because statistics sent from the regency to the provincial office were not carefully checked. Statistical reports were essentially just outcomes of formal and routine tasks. Interestingly, whatever the quality of information or data that were sent to the central government, they had never been refused or queried. A local government official said that "*Everything sent to the central office is always accepted (*laku*), although the statistics have been collected hastily (*ngawur*) and the reports were inappropriately completed.*"[26]

Another problem in implementation was related to the long delay in the budget approval by the central government. This took about 2 years after the submission of the budget proposal to the Ministry of Finance. The delay severely restricted the amount of time available for the project leader to complete the projects in the region and often led to hasty project implementation. In South Kalimantan, another problem resulting from delayed budget approval was that more than two billion rupiah of the 1984 budget for the Department of Transmigration could not be spent. The Governor of South Kalimantan explained that this was partly caused by the interdependencies between the sectoral agencies in implementing the program. In such a situation, although the budget for the Department of Transmigration had been approved, they could not, for example, build houses for the transmigrants because the Office of Land Use in the province was still waiting for their budget approval to be able to survey the land for transmigration (*Kompas,* March 4, 1985). Problems of implementation during the third plan resulted mostly from the separation of DITADA—which was in charge of the selection and planning of sites, land clearing, and construction of infrastructure—from the Directorate General of Transmigration, which was responsible for housing construction, recruitment of transmigrants at the place of origin, and their transportation, resettlement, and provision with 5 years of economic and social support. DITADA was concerned only with the completion of land and resettlement preparation and reported directly to the Ministry of Public Works. Its tasks were mainly related to the physical works, such as land clearing and infrastructure construction, and were relatively less complicated than those of the Directorate General of Transmigration, which mostly dealt with people. As a result of task distinctions between DITADA and the Directorate General of

[25]Interview with Anwar Rahman, the Head of Social and Cultural Section of *Bappeda* in Pekanbaru, January 1988.

[26]Interview, December 1987. The problems of inaccuracies and inconsistencies of government statistics are unavoidable for anyone studying transmigration.

Transmigration, preparation of the physical settlement often was completed although the transmigrants failed to arrive on schedule.

The perception and view of transmigrants toward the implementation of transmigration policy certainly varied according to their personal experiences. Didik Mulyadi, a transmigrant leader who was appointed an acting village head in Pasir Pangarayan, Bengkalis District, told the author that transmigrants most likely will survive in their new settlements if at the time of their arrival the land is in the condition of "*siap-tanam*" (ready to be planted).[27] According to Mulyadi, the span of time in which the land is in this condition is indeed very crucial, because if the transmigrants arrived either before or after this period they would probably encounter problems, for example, weeds growing wild on their land. He remembered when he first arrived in 1980 that the land was in a relatively good condition and ready to be planted. At that time, there were 510 families from Java and 60 families of APPDT. After arrival, they received food rations (*jadup = jatah hidup*) for 2 years, but since their land still had not produced sufficiently, they received additional food from WFP (World Food Program). About ten families left the village just after the food ration from the Transmigration Department was finished, but then around 30 families arrived and resettled as spontaneous transmigrants. In Riau, Pasir Pangarayan has been considered by many people, for example, bureaucrats, academics, and others with whom the author held discussions, as a relatively successful transmigration settlement. Each transmigrant family, besides their home lot of 0.25 hectares in a ready-to-plant condition and a fully constructed house, received 1.0 hectares of land (referred to as *Lahan* I or Land One) from which vegetation had been completely removed. In addition, each transmigrant family in theory should also receive 0.75 hectares (known as *Lahan* II or Land Two) of uncleared land. According to an informant, the transmigrants in his village who received *Lahan* II found that the distance from the village to that land varies, for example, some are only 1 km, but some are about 13 km away. About 60 families, however, have not received *Lahan* II land that was supposed to be provided for them. There was a rumor that those lands were still in dispute because the local people claimed the land as their own.

Problems with local people regarding the acquisition of land for transmigration settlement are relatively rare, either in Riau or in South Kalimantan. This is not surprising since much land in these provinces is unfertile and therefore not being used by the indigenous people. In many cases, disputes over land with local people are related to the land allocation for PIR-Trans projects, in which the Transmigration Department only deals with the recruitment of transmigrants, whereas the process of land acquisition from the locals is handled directly by PTP and *Dinas Perkebunan*. An example of such a dispute occurred in the district of Kampar in Riau in which 550 family heads from several villages in Siak Hulu subdistrict complained that they had been pressed by the PTP to give up their land for PIR areas without any compensation. They brought this case to the DPRD (Provincial People

[27] Interview, August 10, 1986.

Representative Council), but they failed in their attempt to talk with the governor. According to the Bupati of Kampar, he has explained to the people that they would not receive compensation in the form of money since as the landowners they would be given priority to be PIR transmigrants (*Kompas*, February 23, 1985). It was also reported that some local people rejected the offer to become PIR transmigrants because they would have to wait for at least 5 years before the rubber can be tapped. "In the meantime, how shall we make a living?" asked one of these local people. A similar situation was also experienced by local people in Dusun Tengah subdistrict, Barito Selatan District in South Kalimantan. Only 27 out of 299 families whose lands were requisitioned to be cleared for PIR-Trans projects decided to join the program. The majority of these indigenous Dayak people refused to join the program or to reside in the houses, which had been provided for them. As with the local people in Riau, these Dayaks also considered that their economic situation would be worse if they engaged in a PIR-Trans project (*Kompas,* December 27, 1987).

The *Kompas* correspondent in Riau, in a conversation with the author (February 28, 1988), observed that in many cases, development projects such as the PIR-Trans schemes more often disadvantaged the local populations rather than improved their economic life. For example, a local man who owned six hectares "rubber plantation" (*karet rakyat*) could employ around ten workers. Therefore, in addition to earning an adequate income from his land, he also enjoyed a certain social status as an owner and employer. If he joined the PIR-Trans, although he would receive two hectares of land, he would have to give up his six hectares, and his status as an employer would drop to the status of laborer for the PIR Company. More importantly, his economic condition would also be unsure for several years to come because while his own plantation had been cleared away, he would have to wait until the rubber plantation began producing for the first time. An informal local leader in Taluk Kuantan, during an interview with the author (December 28, 1987), also argued that even though the local people have been given their share in the PIR-Trans scheme, they were not usually consulted in advance about the use of their areas by the government for such projects. He observed that in many cases the government ignored or did not pay attention to the local interests or local aspirations. In other words, the local people were treated merely as objects rather than as human beings. It is therefore no surprise that the indigenous people often refused the government plan with conflict occurring as a result. In such situations, the government blamed the local people for being opposed to regional development. But what was happening, he explained, was that the government basically adopted "a wrong approach" in relation to the local people.

Institutionally, the difficulties and problems of implementing the central government plan for transmigration were noted by Tjokroamidjojo (1979, 80) in his Way Abung study:

> *While problems of implementation thus stemmed to a certain degree from the weakness of coordination, they also resulted from other causes, some of which are indicated here. In planning, more emphasis had been placed on the relief of demographic pressures in Java than on regional development and social improvement in Sumatra. Delays in the processing of government financial aid and disbursements contributed to a lowered capacity of the program to absorb transmigrants in destination areas. Government clearance and land,*

lagged behind the need for allocation of plots to newly arrived settlers, and the latter did not have the necessary equipment to clear the land themselves. In addition, official administration of land titles was unduly cumbersome.

For the fourth plan period, the government decided to establish a new Department of Transmigration. It was essentially a merger of the former DITADA of Public Works and the Directorate General of Transmigration. Following the decision to establish a separate department, the personnel of some sections in DITADA were moved to the Directorate General of Preparation and Settlement (PANKIM) and personnel of Directorate General of Transmigration joined the Directorate General of Mobilization and Development (RAHBIN) of the new Department of Transmigration. Although the implementation problems that resulted from the physical separation between DITADA and the Directorate General of Transmigration seemed to have been solved with this unification, the different experiences of the personnel of PANKIM and RAHBIN could not be easily coordinated. The personnel of the former DITADA, which had been dominated by planners, mostly from engineering backgrounds, continued to operate in their own style and tended to complete their tasks faster than their counterparts from RAHBIN. Partly because their tasks mostly dealt with physical activities involving heavy equipment, PANKIM was used to obtain a larger budget from the central government compared with RAHBIN. It was partly because of PANKIM's abundant financial resources that it was able to achieve its target for physical settlement preparation, particularly during the third and fourth plan periods. Preparation of the settlements had to be completed before the RAHBIN staff could commence their work. This situation, according to a senior member of staff in the Department of Transmigration, was the major cause of the large number of unoccupied houses in the project areas, a problem that was given wide publicity by the press in November 1987 and known as the *puso* issue.[28]

Due to the serious problems that the transmigration program had experienced during implementation, the president on October 16, 1985, promulgated an instruction (Keppres No. 59) that the governors in each receiving province should be responsible for coordination of the transmigration program in their area. However, this move failed to solve the problem, mainly because frequent interventions from ministerial offices in Jakarta had more direct influence on program implementation than did the governor. The formal authority of the governor was not strong enough to overcome this direct intervention from Jakarta. Governors have directed their attention to transmigration implementation problems in their provinces, but the effects have often been somewhat token. For example, partly due to strong pressure from

[28] *Puso* is a term that is used to denote food crops that failed to produce, usually because of insect infestation or natural disaster. In the context of transmigration, *puso* was used to denote the unused houses. These houses that mostly needed reconstruction had been completed by PANKIM before RAHBIN had begun to recruit, transport, and resettle transmigrants and just at the time when the budget was experiencing a severe cutback, according to an interview with a senior staff member in the Department of Transmigration, December 1987. The current Minister for Transmigration, Soegiarto, gave a similar reason in his explanation to the press on the issue of *puso* (Kompas, February 28, 1989).

the central government, as well as national press coverage on program delays, the Governor of Riau immediately visited several transmigration projects and ordered the contractors to speed up their work (*Kompas,* January 23, 1980). The governor has less power than the ministerial office in Jakarta in relation to the transmigration program because transmigration has a national priority and is considered as a strategic program. The minister who has direct command over the program's resources and manpower has more power to deal with it and is located in Jakarta, close to the center of power. President Suharto himself has given considerable personal attention to transmigration. This explains why the latter was so influential, and by the same token, it explains the lesser power of the governors during the program implementation. One former director of the Department of Transmigration reported that, in his personal experience, because transmigration had been considered a strategic program, it had been quite easy for government agencies working for the program to request budget funds, no matter how much they needed (Gunawan 1986, 5).

In the fourth development plan, partly because the coordination between agencies had created many problems during the implementation of the third plan, the president decided to establish a separate department that dealt mainly with transmigration. The new department was essentially a merger of DITADA of Department of Public Works and the former Directorate General of Transmigration. As a merger of two different departments, adaptation and consolidation were required. These proved difficult. Although transmigration was to be carried out by a single department, coordination remained a serious problem, in particular between the two major directorate generals of Settlement Preparation (PANKIM) and Mobilization and Development (RAHBIN). The two directorates have continued to act independently in practice partly due to their different backgrounds and previous experiences. Since the end of 1986, the governors have been given formal authority to coordinate the transmigration program, but in reality, have little power and are not influential. Within the government structure in Indonesia, the governor and his administration are just another sectoral agency in the region, subordinated to the Ministry of Home Affairs. Regarding transmigration, the governor's problem is even more difficult since transmigration is a national priority in which President Suharto himself takes an active personal interest. It is not an exaggeration to suggest that transmigration had become a sacred program, an article of faith that no one dared to oppose. In such circumstances, as precisely argued by Arndt (1984, 49), "*to recommend the abandonment of the transmigration policy, however sound the arguments might be, would be pointless.*" Another major problem during the program implementation was related to the availability of funds. Since the transmigration program is almost totally dependent on the central government budget, a major cutback since 1986 has drastically affected the program implementation.

Many of the problems experienced in the implementation of the transmigration program were basically a result of the joint connection of a top-down and centralized approach among the implementing agencies at the center, and a strong, centrally imposed target orientation among the implementing agencies in the regions. In the boom periods, starting from the beginning of the third plan until the middle of the fourth plan, as the transmigration program became an important vehicle for developing the regions, the implementation of the program become an arena in which each agency attempted to achieve its own physical target. This was especially the case for the implementing agencies under the Ministry of Public Works. However, integration with other agencies was often neglected. Coordination, particularly in the third plan period, was essential due to the many different departments that dealt with transmigration. However, effective coordination was rarely achieved. In the regions, disputes among the implementing agencies involved with the transmigration program were almost unavoidable because each agency, including Department of Public Works, Agriculture, or Transmigration, acted on the directions of its office in Jakarta. When the state apparatus itself monopolizes economic and social power in the society and retains full discretion over policy initiatives, the system can be labeled closed. Its policies generally respond to the institutional interests of the group that dominates the government machinery. Commonly in such systems, national elites conclude that their goals for the society, such as economic development or national security, are prejudiced by the social and economic patterns characterizing "marginal populations" (Grindle 1980, 284).

In Indonesia, since there is no "democratic" avenue for the civil society to express or to articulate its views, it is difficult to judge whether a government policy is genuinely supported by the majority. In such an environment, the real problem is the lack of viable control and feedback from society. Transmigration implementing agencies and their bureaucrats at the operational level were mainly motivated by the urgency to achieve the physical and measurable targets, such as the number of houses to be built, the length of roads to be constructed, or the number of families to be moved. In addition, the hierarchical structure of the transmigration implementing agencies created a bureaucratic attitude that is popularly known in Indonesia as "so long as the boss is pleased" (ABS). This attitude was adopted by subordinates, especially in relation to their direct superior in the office, in which they would perform or act first and foremost to fulfill the interests of their superiors. This kind of attitude, among others, often resulted at the operational level in neglect by the implementing agencies of the needs of transmigrants or local people.

Observations and interviews in Riau and South Kalimantan suggested a certain level of rejection of the program among people in the region because it was strongly imposed by the central government and often ignored the real needs of the local people. Again, given the weak bargaining power of the local people in relation to the central government, their dissatisfaction was quietly suppressed. The transmigration policymaking process has been almost totally dominated by the central government and allows only a minor role for the regional government. Within the structural organization of the Indonesian government, the role of the provincial-level government is hierarchically subordinated to the central government. In the policymaking process, the regional government's primary task is to provide information to the planners and decision makers in the central government. In the implementation of the plan, the provincial-level governments have played an important role, particularly in implementing the central plan and decisions. However, their activities were mostly contained within guidelines determined by the central government.

Transmigration can be classified as a broad and complex policy with a large formal structure of organization. Since 1983, it has been an independent department.

Conclusion: Transmigration, an Ideological Policy

As a major part of transmigration policy involves the distribution of land, shelter, and other facilities, it can be described as a distributive policy. However, the main characteristics of transmigration policy resemble the type of program that has been labeled "ideological". According to Quick (1980, 42), there are three main characteristics of an ideological program. The first characteristic is that the program is expected to realize a multitude of goals simultaneously. The program represents a working model of the new society and is therefore expected to introduce changes in many aspects of existing social relationships. In addition to being numerous, the goals set for an ideological program are also ambiguous in that national elites rarely have a clearly worked-out view of what needs to be moved from the existing state of affairs to the new one. The second characteristic is that the program has no hierarchy of goals, that is, no clear statement that goal A is more important than goal B, or at least that goal A should be achieved before embarking on goal B. The ambiguity of goals presents the implementing organization with the clear message that everything must be done at once and that there are no priorities that can be used to orient implementation. The third characteristic of the goal structure of ideological programs is the immeasurability of many of their objectives. The target population is expected to live differently, to act and think differently, and to progress toward many of those goals, but such objectives cannot be measured with any precision. Agencies charged with implementing ideological programs thus have a goal structure that consists of a few measurable goals and many nonmeasurable one.

Ideological programs are also affected by the high expectations of the national political elite. Such programs are endowed with a sense of urgency and vital importance, placing the implementing agency under extreme political pressure to produce results. So important is success in these programs those political appointees are selected to head the implementing agency and resources are made available to this agency that are unavailable to other programs. The organizations charged with implementing ideological programs are also relatively immune from criticism, shrouded as they are by the national priority of their program. These, then, are the basic characteristics that define an "ideological program": multiple, ambiguous, and nonmeasurable goals; high expectations; resource availability; politicized leadership of the implementing organization; and immunity from public criticism.[29] The main

[29]The immunity of transmigration policy from "independent" evaluation has been admitted, for instance, by a *Bappenas* official on one occasion where a closed meeting was held between *Bappenas* and the Centre for Policy Implementation Studies (CPIS). CPIS is a "semi-government" consulting agency, attached to the Ministry of Finance, and assisted by Harvard Institute for International Development (HIID). A major task of CPIS is to advise the Indonesian government,

characteristics of transmigration policy, such as its multiple, ambiguous, and nonmeasurable goals, resource availability, immunity from criticism, strong political support from the ruling elites, and high expectations, suggest that transmigration is an ideological policy. As an ideological policy, transmigration produces formidable counterproductive forces. For example, its immunity to public criticism has created an implementing agency that lacks viable control and feedback. Such counterproductive forces have, in turn, hindered the implementation process.

The target groups, which consist of potential transmigrants, transmigrants, and indigenous populations in the receiving areas, are people with only limited bargaining power and alternatives within the existing social, economic, and political environment. Their powerless position within the implementation process has allowed the implementing agency to treat them as mere objects rather than subjects. The evidence from Riau and South Kalimantan suggests that the context of central–regional relationships has played a crucial role in the implementation of transmigration policy. The position of the provincial-level government, as well as of the local society at large, was subordinated to the interests of the central government. They were thus structurally weak in relation to the central government. Because of this unequal relationship, the implementation of transmigration policy has suffered from the lack of important controls and feedback mechanisms that could transform it into a viable and implementable policy. Analyzing transmigration policy within the wider context of center and regional relationships indicates that transmigration policy clearly reflects the strength of centralization and the primacy of national government in Indonesia's development.

---

particularly through the Minister of Finance, on the implementation of various government projects. The advice was sought after an evaluation of a certain program was conducted by CPIS. In the meeting, the official from *Bappenas* warned the CPIS not to be involved in evaluating two government programs, transmigration, and cooperatives. The reason given was because those two programs were considered as "too political," and CPIS would face many problems when they attempted to evaluate those two departments. (This information was obtained from a personal communication with a CPIS staff member in January 1988.)

Part II
Migration, Development, and Centralized State

MIGRATION IS AN INTEGRAL part of the development process, both caused by changes in specific characteristics of economies, societies, and populations: and its causing changes in those economies, societies, and populations. In this regard, the gap in the growth of the working age population, for instance, particularly between Java and the Outer Island—a result of historical lower fertility and the earlier start of the demographic transition in Java—should be specifically considered as migration that is generally concerned with the productive age group. While many studies about migration in Indonesia are mostly confined to the demographic and economic aspects, this part is an attempt to look beyond the usual conventional approaches. The unequal power relations between central and provincial governments will be elaborated as a crucial factor in explaining the social and political structures that contextualize migration in Indonesia.

Indonesia—given its ethnic plurality and archipelagic geography—places population mobility as a crucial factor in the making of a modern nation-state. The recent incident of ethnic and religious conflicts resulted in many dislocated populations strongly indicated that primordial tendencies are still intact. In the New Order, as infrastructure for transportation and communications accelerated, movement of people was significantly increased: ensuing a simple "supply and demand" economic dictum. At the same time, people's movement from Java and Bali to the other islands in the archipelago under the state's demographic engineering process, particularly through the transmigration program, also substantially accelerated. The relationship between migrants and the local population is not always in harmony as shown by sporadic incidents of social tensions and conflicts. The previous conflicts, however, have never been turned into an escalated scale that could destabilize social cohesion, as occurred recently. In the aftermath of the economic crisis and the downfall of the New Order's main pillar, Suharto, rampant ethnic and religious conflicts clearly appalling the belief in interethnic and interreligious harmonious life in Indonesia.

This part consists of three chapters, primarily links migration and development in Indonesia, with a special focus on the experiences during the New Order period. The first chapter provides a general overview of the relationship between population mobility and development. As migration is perceived as an integral part of the

Chapter 4
Migration Patterns and Development

Compared to other social science research, the study of migration is relatively new in Indonesia. Probably, the various types of migration and the wide area over which those various types of migration occur are constraints to the study of migration in Indonesia.[1] However, the since early 1950s, there has a growing interest among researchers to study the migration behavior of some ethnic groups. Attention was drawn to migration studies in the 1970s when there was a need among the policymakers to understand the migration process of the Indonesian people. For example, there was a need to understand how to attract people from the overcrowded islands (Java, Bali, and Lombok) to the sparsely populated islands (Sumatra, Kalimantan, Sulawesi, and Irian Jaya). The problems that increased as a result of rural–urban migration such as urban unemployment, housing problems, and rural unbalanced age composition also encouraged policymakers and researchers to investigate the determinants and consequences of population mobility.

Migration Study in Indonesia

Even though the history of population mobility is as old as the history of the human being, the study of migration in Indonesia was not begun until the country was colonized by the Dutch. The Dutch historians, Vollenhoven, and van Leur, as quoted by Hugo (1980, 97), summarized patterns of migration during precolonial Indonesia as being of three main types:

[1] It has been shown that, even though Zelinsky (1971), among others, has defined migration as any permanent movement, in this review the word "migration" is still used as synonymous with the word "population mobility" or "population movement." Furthermore, the review distinguished between permanent migration and temporary migration (circulation).

R. Tirtosudarmo, *From Colonization to Nation-State*,
https://doi.org/10.1007/978-981-16-6437-3_4

1. Colonization by large groups of migrants from one region who settled in another region.
2. Migration of individuals, particularly traders, who settled in port cities.
3. Establishment of authority in foreign regions.

From the sixteenth century up until the latter part of the nineteenth century, there was a gradual increase in the number of Europeans, whose activities had much influence on the pattern of migration in Indonesia. On October 7 and 8, 1930, the Netherlands East Indies administration conducted the first population census. Only a few studies have been done based on the results of this census, but the population mobility information obtained was the most comprehensive about the volume of migration during the colonial period (Hugo 1980, 105). According to this census, 11.5% of the total indigenous population of Indonesia was living outside their district to birth. It was interesting that a large number of Java-born Indonesians were enumerated in the other islands, especially Sumatra. Pelzer (1945a, b, 189) showed that in 1930, 967,000 Javanese, 120,000 Sundanese, 16,000 Madurese, and 39,000 other Indonesians of Java ancestries were found in the Outer Islands: although not all of them were Java-born.

Two factors contributed to the migration from Java to the Outer Islands. First, the increase in the Dutch plantations in the other islands where labor was scarce led planters to recruit people from Java. The first group recruited was Chinese coolies, but later Javanese and Sundanese workers were also recruited. Secondly, in 1905, the Dutch government considered Java to be overpopulated and introduced its colonization program to resettle people from Java in the Outer Islands. However, the concentration of colonial activity in Java also led to several growing urban centers, such as Batavia, Surabaya, and Semarang. These cities attracted people from the crowded interior because of the employment opportunities offered there (Pelzer 1945a, b, 175). Hugo (1980, 114) also showed that migration played a key role in the growth of urban centers during the colonial period. Although a significant proportion of urban dwellers were immigrants from overseas, in-migrants from rural areas and the Outer Islands made up majority among the indigenous urban population. During the colonial period, besides movement due to the impact of colonization, there was also large-scale migration by people from the highly mobile ethnic groups of the Outer Islands: the Minangkabau, Bugis, Banjarese, and Makassarese.

During World War II (1942–1945), Indonesia was occupied by the Japanese. Little is known about population movement because no data were available during this time. The only evidence was that the Japanese initiated a forced labor *(romusha)* recruitment system, which resulted in many Javanese being sent to the Outer Islands or other countries in Southeast Asia; many of them never returned. After independence, the new Indonesian government reinstituted the colonization program under the name of *transmigration*, and the first transmigrants were settled in Lampung (South Sumatra) in 1950. From the early 1950s, Western anthropologists and sociologists paid more attention to the migration of selected Indonesian ethnic groups. Cunningham (1958) studies the mass migration of Bataks from the Toba highlands in North Sumatra to the East Coast in the early 1950s. Brunner (1961) also studied urbanization and ethnic identity in North Sumatra. In the early 1960s, Vredenbreght (1964) studied the movement of Baweanese who have a strong propensity to migrate. Additionally, the Minangkabau of West Sumatra is probably the group, which has been studied in the greatest detail (Naim 1979; Murad 1980).

The principal theme that emerged from these various studies of the mobility of ethnic groups was that the pattern of mobility exhibited was apparently due to a "tradition of migration" or to "traditional pressures" within these ethnic groups. The value of such an explanation to the pattern of migration at the regional and national level, however, must be considered limited. Most studies tend to consider that the forces behind the migration of these ethnic groups are more or less constant. However, Forbes (1981, 61) showed that the forces that produce this pattern were constantly changing. Some forces (e.g., political unrest) were only significant at particular periods, and other forces (e.g., ecological, demographic, economic, educational, urban attractions) increased steadily and significantly, while others (e.g., geographical and social–systemic) remained constant throughout the period or declined significantly declined.

In 1961, the Indonesian government carried out its first population census. McNicoll (1968) provides the most detailed studies of migration from the census. He showed that besides seeking a better life in urban areas, the reason for rural to urban population movement is feeling of insecurity due to political instability in some rural areas (i.e., Darul Islam rebellion in West Java). The second census after independence was conducted in 1971. Two conventional types of migration estimates can be obtained from the 1971 census data. These are referred to, respectively, as lifetime and recent (5 years) migration (Speare 1975). Several studies have been done based on the 1971 census data: Among those are Speare's (1975) study and how to interpret the migration data from the census; and Sundrum's (1976) and Titus' (1978) study on relationship between rural–urban migration and unemployment. Moreover, the study of urbanization in the primate cities, particularly in Jakarta, also received some attention (Castles 1967; Papanek 1975; Temple 1975; Jellinek 1977; Hugo 1978).

Since the mid of 1970s, the study of migration in Indonesia has also broadened significantly to include studies of adjustment of migrants to cities (Suharso 1978; Honggominarso 1977; McCutheon 1978; Pelly 1983), the lives of poor rural migrants in Jakarta (Papanek 1975; Temple 1975; Jellinek 1977), the place of origin of migrants, the determinants of migration, and the impact of migration on fertility (Hugo 1978; Suharso 1978; Koentjaraningrat 1975; Mantra 1981; Oey 1977). However, probably the most important development in migration research in Indonesia, which has occurred during recent years, has been the recognition and increasing attention given to short and nonpermanent movements and commuting (Papanek 1975; Temple 1975; Jellinek 1977; Forbes 1981; Hugo 1978; Koentjaraningrat 1975; Mantra 1981).

Modes and Patterns

The first problem that arises when we discuss migration is the problem of definition. In the case of permanent or temporary migration, the difficulty relates to the time limit upon which to define a movement as permanent or temporary. Zelinsky (1971, 225–226) defined migration as any permanent or semi-permanent change in residence: More meaningfully perhaps, it is a spatial transfer from one social unit or neighborhood to another, which strains or raptures previous social bonds. On the other hand, he indicated circulation as a great variety of movements, usually short-term, repetitive, or cyclical, but all having in common the lack of any declared intention of a permanent or long-lasting change in residence. In Indonesia, the census is the most important source of data on permanent migration. However, censuses or large-scale demographic surveys (i.e., National Sample Survey) usually use criteria and questions designed to detect predominantly longer distance migration. Shorter distance movements, such as circulation, seasonal migration, and commuting, which are usually nonpermanent, are unrecorded and tend to be ignored. This unfortunate situation has resulted in a lack of data at the national level on temporary movement. However, in recent years, there has been increasing attention among researchers to the study of nonpermanent migration in Indonesia using data from smaller surveys.

Based on the 1930 census, three types of migration can be identified (Hugo 1980, 106–121). The first was the migration into plantation areas either in Java or outside Java. At that time a, substantial proportion of the population in the plantation areas was born outside their region. Sugar plantations particularly required many workers. For instance, districts located in sugar plantation areas in Central and East Java attracted many migrants. By the late nineteenth century, the growing congestion in Java led planters to recruit labor from Java to be sent to the Outer Islands. Next, the second type was migration to urban areas. The expansion of Dutch economic activities assured the supremacy of the coastal cities, which functioned as trans-shipping ports for the agricultural products of their hinterland. In all provinces, urban areas constituted major destinations for in-migrants. Lastly, the third type was international migration. During the colonial period, there was the influx of a European ruling colonial elite and large numbers of Chinese and to a much lesser extent, Indians, and Arabs. Emigration of indigenous Indonesians occurred, but it is only on an exceedingly small scale: Most of the Javanese are under contract coolie recruitment programs to obtain cheap labor for plantations in Malaya, Suriname, and New Caledonia.

In contrast with the 1930 census, the 1961 census used the province as a space unit with a time reference of 3 months to measure migration. Migrants were defined as those who crossed the province boundary and had been away for at least 3 months. Based on the 1961 census birthplace data and the 1964–1965 National Sample Survey, McNicoll (1968, 37) found that South Sulawesi showed the greatest internal mobility, followed by North Sumatra and South Sumatra. The last two regions were also prime recipients of migrants. Regions of particular stability were West Nusa

Tenggara and Bali, while Central Sumatra was important as a region of out-migration only. The proportions of out-migrants from West and Central Java were comparatively large, due to urban migration to Jakarta (McNicoll 1968, 33). In 1961, rural Sumatra, Kalimantan, and Maluku contained the largest "external" population. In Sumatra and Kalimantan, the in-migrants were transmigrants mainly from Java, while in Maluku, the majority was from Sulawesi. At the other extreme, the rural areas showing the fewest in-migrants relative to their population and the least internal mobility were in the province of Java, particularly East and Central Java, Nusa Tenggara, and Central Sulawesi. Similar to the 1961 census, the 1971 census used the province as a spatial unit, but with 6 months as a time reference. According to the 1971 census, nearly 5% of the total population had migrated from their province of birth and another 1.2% had changed their province of residence more than once (Sundrum 1976, 89).

The most important stream of migration (nearly 40% of the total net lifetime migration) was that flowing into Jakarta from all other provinces. About four-fifths of these migrations were from other provinces in Java. Although the population of Jakarta was one-third of the urban population of Java, about two-thirds of all migrants to urban areas ended up in Jakarta (Sundrum 1976, 90). The second largest stream of migration was that flowing into Lampung, Jambi, and Riau. The percentage born outside those provinces was 36, 16, and 13, respectively (Speare 1975, 66). The rest of the interprovincial migration was to provinces experiencing high rates of economic development in recent times, such as East Kalimantan where there were timber and oil industries. Furthermore, Speare (1975, 79) indicated that interprovincial migrants are more likely to settle in urban areas than in rural areas. Fifty-five percent of all migrants between 1966 and 1971 were living in urban areas in 1971. Considering that most of the migration studies that used the census data dealt mainly with the volume and streams interregional migration, the National Institute of Economic and Social Research–Indonesian Institute of Sciences (Leknas-LIPI) in 1978 conducted a large sample survey on rural–urban migration (Suharso et al. 1976). This survey was carried out in ten provinces to study the characteristics, motivations, and success of migrants in the cities. According to this survey, the net movement of people in Indonesia during the 1960s from rural to urban areas was not very great except for the movement into Jakarta, which grew very rapidly (Mantra 1981, 4). Although for most cities net migration was either small or even negative, many experienced a considerable total flow of people in and out. One apparent reason for the back-and-forth movement between villages, towns, and cities was the need to go to the urban centers to obtain secondary and higher education. There was also temporary migration to the cities during the slack period in farming.

Most of the studies on circulation in Indonesia were conducted on a microlevel or regional basis. Some of these studies were concerned mainly with the highly mobile ethnic groups. Forbes (1981), Hugo (1978), and Jellinek (1977) studied circulation in a particular province or city, while Mantra (1981) conducted a detailed study of circulation between two *dukuh* (subvillage) in Central Java. Little is known about circulation during the colonial period and the years after independence. The unavailability of data on short-term migration during those times was probably a major constraint to studies of circulation. Hugo (1982, 60) found that according to the 1971 census some 7.3 million Indonesians, or 6.4% of the population, had lived at some time outside their province of present residence: Hence, they were classified as "migrants." However, the criteria used to define migrants in the census excluded most short-distance and short-term movers. Community-based studies in Jakarta have pointed clearly to the importance of nonpermanent migrants in that city (Temple 1975; Jellinek 1977). A study of the nonpermanent movement, which was conducted by the Environment and Settlement Studies Center—Jakarta Special Region in 1979, showed that there were 51,331 circular migrants in 1533 *pondoks* (rooming house). Most of those migrants worked in the informal sector. There were around 200 kinds of informal sector jobs, which were scattered in all areas of Jakarta. The migrants usually go back to their village three times a year. Most of them own land and a house in the village.

Jellinek (1977, 67), in her study of petty traders, found that of more than 200 mobile traders, which were interviewed, all but one was circular migrants. She suggests that if her study is representative of mobile traders in Jakarta, then there must be many circular migrants engaged in petty trading alone: in addition to those working as day laborers, pedicab drivers, and the like. According to Jellinek (1977, 690), the frequency with which migrants move back and forth depends to a great extent on the distance of their village. Migrants from West Java generally return once or twice a month, while those from Central Java generally return once 3 or 4 months. Temple (1975, 78) showed that only 25% of migrants who had been in the city for 6 months or less planned to stay permanently. Furthermore, he said that there was a seasonal pattern in the flow of migrants to Jakarta, with December and January as favored months. Temple (1975, 79) also showed that the main reasons given by the migrants for leaving the village were related to the lack of employment in the rural areas. The traditional pull factors, such as the possibility of higher paid jobs, the excitement of city life, and new educational opportunities, seem less important. Jakarta was chosen because migrants felt it offered them the best hope of finding reliable employment opportunities.

Koentjaraningrat (1975, 113), in his study of population mobility in the rural fringe of Jakarta, proved that, basically, most people were not willing to migrate when work could be found in their village. People from two villages studied around Jakarta did not migrate solely for reasons of lack of employment opportunities in their villages, but rather to seek additional sources of income during the favorable seasons. Commenting on the preliminary result of the 1980 population census, Hugo (1982, 63) indicated that "Jakarta's population was 6.5 million, representing an annual growth rate of 4%. This was somewhat lower than the 4.4% during 1961–1971 and certainly lower than most commentators predicted. Undoubtedly this growth rate severely underestimated the growth of Jakarta's functional population: the census criteria referred to earlier would have ensured that most of their income in Jakarta were not enumerated in that city but their village of origin."

In 1975, Mantra (1981) conducted an intensive study in two *dukuh* (hamlet or subvillage) near Jogjakarta in Central Java, for an attempt to prove that the Javanese is not an immobile ethnic group as people always said. In his study, he discovered three types of population movements that were practiced among the Javanese villagers: commuting, circulation, and migration. Commuting and circulation are common occurrences and are not viewed as unusual actions. According to Mantra (1981, 169), the economic needs in those two *dukuh* underlie population movement but did not fully explain it. Social and kinship ties, the desire for further education, and the perception of opportunities at the other destinations often played an integral part in the decision-making process whether to move or to stay. This substantiates the basic proposition that economic and social factors, in combination, explain movement away from the village and the return to it.

The most comprehensive study about voluntary migration of highly mobile ethnic groups is probably the work of Naim (1979). He made an intensive study about why and where the Minangkabau migrate (*merantau*), and the consequences upon the place of origin and the place of destination. According to Naim, *Merantau* as a type of migration behavior is unique. It is a voluntary movement, usually leaving one's cultural territory whether for a short or long time, with the aim of earn a living or seek further knowledge or experience. In the migration literature, *merantau* is considered as temporary migration or circulation because the migrants have the intention of returning home. Naim (1979, 51–56) compared the migration volume of some of the most highly mobile ethnic groups since 1930. In 1930, the percentage of migrants among the Bawean was the largest (35.9%), followed by the Bataks (15.3%), the Banjarese (14.2%), and the Minangkabau (11%). The projected migrant population in 1961 showed that the Bawean is still the most mobile (31.6%), followed by the Minangkabau (31.6%), the Bataks (19.5%), and the Banjarese (12.22%). In 1971, the projected Minangkabau migrants were 44%, probably the highest among the ethnic groups in Indonesia.

The island of Sulawesi is also the homeland of several of Indonesia's most peripatetic ethnic groups. The Bugis, the Makassarese, and the Torajan are the three largest ethnic groups in the province of South Sulawesi. Lineton (1975a, b) showed that out-migration has a long history among these ethnic groups. The fact that the Bugis have been migrating for centuries, in itself, is a factor predisposing other Bugis to leave their native land since the Bugis colonies abroad are a source of information about more favorable economic opportunities elsewhere and often provide both financial assistance and moral assistance to the would-be migrant. The success of many former migrants arouses a spirit of emulation in those who have remained behind. Migration is seen to be a means, for many people the only means, to achieve wealth and a higher social status (Lineton 1975a, b, 190–191). There is also substantial Bugis movement within South Sulawesi, including seasonal circular migration between the village and the provincial capital city of Ujung Pandang. Much of this rural–urban movement also has a seasonal rhythm and involves the Makassarese and the Torajans (Forbes 1981).

Another mobile ethnic group is the Bawean people. Bawean is the name of a small island in the Java Sea. Vredenbreght (1964) showed that *merantau* among the Bawean has a deep cultural root. The island of Bawean, sometimes known as the "island of women" on account of its great preponderance of females, was long characterized by decreasing population as the result of male migration (1964, 113). The circular aspect of their migration is freely expressed by the Bawean in conversation, for instance: "We have not come to Singapore like the Chinese, to settle and grow rich, but only to earn money and buy clothes, and once we are old, to return to Bawean for good" (Vredenbreght 1964, 128). Many of the Bawean ethnic groups live in Singapore and Malaysia, mainly concentrated in Penang, Selangor, and Johor. Up to now, little is known about nonpermanent population mobility in other provinces. However, Hugo in his recent article (1982) mentioned that research had been done on circular migration in several provinces in Indonesia, such as Aceh, Irian Jaya (Rumbiak 1984), and South Kalimantan.

Determinants and Consequences

From the colonial period until recent years, foreign economic influence plays a key role in population mobility in Indonesia. This influence upon the labor movement was clear during the Dutch colonization period (Pelzer 1945a, b; Geertz 1963; Penders 1977; Hugo 1980). The Dutch dominations and penetrations of capitalistic production-oriented to world trade reshaped the society and the economy of Indonesia and brought about major changes in the pattern of population mobility. After independence, the influence of foreign economic activities on the general development strategy was obvious during the Suharto Era. During the past 20 years, the Indonesian government has a planned economic development program to get the country moving ahead. Since 1967, the Indonesian government introduced the open the door policy and allowed foreign investors to operate in the Indonesian economy.

Titus (1978) using Amin's center–periphery approach (1974) has tried to show that development strategy and foreign economic activities play an important role in affecting interregional population movement in Indonesia. The most favored foreign company investments are the extractive industries, such as timber and oil industries, as well as assembling and construction industries, followed by commerce and service industries. Those kinds of activities have attracted a great deal of foreign capital. Furthermore, the development of the extractive activities is almost exclusively limited to the Outer Islands (McCawley 1981), whereas the other activities are mainly conducted in the urban centers on Java. The capital city of Jakarta alone has attracted 50% of all foreign investments excluding the extractive sectors (Sethuraman 1976, 27). Titus (1978) classified the province as a center or periphery based on the presence of such economic activities. In addition, every region (province) was assessed based on the existence of "attraction factors" (i.e., employment opportunities, regional per capita income, urban attraction) or "expulsion factors" (i.e., rural population pressure, ecosystem instability, education factors, political upheavals) (Table 4.1).

Titus (1978) also showed that the greatest mobility together with net immigration is to be found in the economic "boom" provinces of both the center type (e.g., Jakarta, North Sumatra) and the relatively developed periphery type (e.g., South

Table 4.1 Indicators of Indonesian Economic Development, 1965–1990

Indicators	Mid-1960s	c.1990
Real GDP per capita	1960–1965	1988–1990
Growth (%)	−0.4	±5
1990/$	190 (1965)	570 (1990)
Share of GDP (%)		
Agriculture	50	22
Industry	13	40
Manufacturing	8	20
Industry: % of food and rubber products in total manufacturing	60 (1963)	12 (1986)
Industrial production per capita:		
Textile (meters)	4.1	28.0
Electricity (kWh)	17.7	194.5
Fertilizer (kg)	1.1	39.1
Agriculture:		
Rice consumption/capital (kg/year)	92	155
Rice yields (Java only, tons/ha)	1.7	3.2
Investment:		
Gross domestic investment as % of		
GDP	8	36
Inflation (% increase)	>500	8–10
Debt.		
Total ($ billions)	2.4	68
% of exports	524	229
Transport/infrastructure:		
Domestic air travel (million departures)	0.4	9.0
% of roads in good condition	5	31
Registered motor vehicles ('000)		
Buses and "colts"	20	469
Trucks	93	1.024
Motorcycles	308	6.083
Poverty:		
Java % very poor	61	10
% sufficient	8	36
Outside Java % very poor	52	7
% sufficient	10	47

Source: Hill (1994, 125)

Sumatra, Riau, East Kalimantan). The lowest mobility and a zero-migration balance are to be found in the isolated and still largely self-sufficient periphery type of province, i.e., East and West Nusa Tenggara. Meanwhile, the highest mobility together with net out-migration appears in the highly integrated but stagnating peripheral provinces close to center groups (i.e., West Sumatra, Central Java, and Jogjakarta). The result of Titus' study on interregional migration is similar to the pattern of interprovincial migration streams, which were shown by Speare (1975) and Sundrum (1976) based on the 1971 census. Criticizing the work of Titus, which ignored short-term and nonpermanent movements, Forbes (1981), using the same approach, conducted a study in South Sulawesi, mainly Ujung Pandang. He focused his attention on the circulation of petty traders between Ujung Pandang and the hinterland. Forbes (1981, 65) showed that around half and as many as two-thirds of migrant petty producers kept in contact with their villages with regular trips back. There is an important circulation between Ujung Pandang and the hinterland not only of labor but also of cash and goods. In this manner, the main theme of Forbes' study is the importance of viewing mobility as a reflection of the structure and the processes of Indonesian society.

The decreased availability of land can theoretically be offset by investing in technology or intensifying cultivation. Unfortunately, the modernization of agricultural production in some rural areas in Java has created more unemployment (Montgomery 1975; White 1976; Singarimbun and Penny 1976; Sinaga 1978). For example, in rural Java, the use of the huller and small tractors has drastically reduced the number of farmworkers. There has been a shift of labor from agricultural sectors to nonagricultural sectors, mainly in urban areas, such as domestic services or petty trade (Temple 1975; Suharso, et al. 1976; Jellinek 1977; Hugo 1978; Anaf 1984; Tirtosudarmo 1984). Moreover, in his study, Montgomery (1975) showed that new rice strains did not have any effect on agricultural employment because they were only meant to make Java self-sufficient in rice.

In this part, we turn our attention to one of the most important components of all disparities in income distribution, namely the disparity between urban and rural areas. According to Booth and Sundrum (1981, 202), by 1976 the rural–urban disparity in income had increased, especially in Java. The urban–rural disparity in nonfood expenditure is much greater than in food expenditure. Within food items, there is truly little disparity in expenditure on "basic foods," such as cereals and cassava. The disparity is greater to the "medium foods," such as fruits, vegetables, and tobacco, and highest for the superior food as meat, fish, eggs, and milk. Among nonfood items in 1970, the disparity in housing expenditure was quite low, whereas it was remarkably high for miscellaneous items, which include consumer durables. In all cases, the urban–rural disparity was higher in Java than in Outer Islands (Booth and Sundrum 1981, 203). From 1970 to 1976, because of the government monetary stabilization policy and the concentration of development areas, urban incomes have increased much faster, especially in Jakarta and other large cities than in rural areas. Such changes would normally have led to migration from rural to urban areas (see Steele 1980; Wirosuhardjo 1984).

There is no doubt that during the Suharto Era, many improvements have been made in almost every aspect of social life in Indonesia. During the first five-year plan (1969–1974), the government emphasized the development of infrastructure and communication, i.e., roads, railroads, harbors, and airfields. In the social sectors, the growth in education was also dramatic. While only about half the children at primary school age were enrolled in schools 15 years ago, now many more children at that age have access to schooling. Each village with a population of around 2000–3000 people has at least one primary school and many have more than two (Wirosuhardjo 1980, 3). Health facilities have entered the 50,000 villages with the provision of integrated health services for every subdistrict. At the beginning of the second half of the 1970s, satellite communication was introduced. Later, the government distributed television sets in every subdistrict. The combination of the increase in education, health services availability of transportation, and communication facilities has caused rising hopes and aspirations of rural young people, which resulted in the movement to urban centers.

Tirtosudarmo (1984), in a study on East Java, concluded that the various structural changes in East Java during the last 30 years and creating a dichotomy between the rural "push" and the urban "pull" at the societal level also affected the individual, either directly or indirectly, by changing the costs and benefits of migration. For example, the young, educated, and unmarried population would be more likely to migrate than their older, less educated, and married counterparts, primarily because the former were able to adapt easily to the new situations that resulted from structural changes in the society. In addition, it is likely that the personality characteristics of the migrants, such as more adventurous behavior, willingness to take a risk, and ease of adjustment, were also important factors in migration decision making. Finally, the presence of relatives and friends in urban destinations played a critical role in the migration decision, especially in mediating information about the urban destination and providing assistance for migrants in the new place of destinations.

Based on data for the 1970s, Speare (1975), Sundrum (1976), Suharso et al. (1976), Hugo (1978), and Suharso (1978) came up with the same conclusion about the age characteristics of migrants. Migrants were concentrated in the younger age, particularly 15–24 years of age. Migrants to the smaller cities tended to be younger than those to the large cities, presumably because more migrants to the small cities were coming for educational purposes (Suharso et al. 1976). Sundrum (1976, 86) indicated that return migrants had a higher age group than lifetime migrants. The difference between the average ages of return migrants and lifetime migrants is larger in Java than in Sumatra. This situation suggests that those migrants who return to their original province do so quite soon after their out-migration, perhaps because of difficulties experienced in their new settlements.

According to the Leknas Migration Survey (Suharso et al. 1976), more than one-half of migrants were not married at the time of the survey, and since more had married between the time of migration and the time of the survey, the proportion of who were not married at the time moving was probably much higher. Most of the migrants had not worked before moving, primarily because many were still going to school up to the time when they moved. Among the female migrants, about one-third were married before moving. The education characteristics of migrants to Jakarta and other provinces of Java show a higher educational level than native-born residents in these provinces. The reverse case happened in Sumatra and other islands. This distribution may reflect the fact that more educated migrants, who are also younger persons, have been involved to a considerable extent in the migration to Java. While the migration to Sumatra and other islands consisted to a greater extent of the movement of unskilled workers: presumably, the migrants under "colonization" and "transmigration" program.

The most common occupation for those who were employed was farming, but this accounted for only 12% of the total migrants. Although about 75% of the migrants come from rural areas, most were either out of the labor force or employed in nonfarm work before moving (Suharso et al. 1976). At the village level, Mantra (1981, 151–154) showed that more males than females wanted to migrate from the *dukuh*. Married women did not want to leave their local community, especially since their attention was generally focused upon the children, close kin, and proximate neighbors. The young educated and unmarried people from rural and small cities were groups who were pulled by the extension of job opportunities in the urban centers. In addition, the decreasing job availability in agricultural sectors has pushed them to seek a job in the cities. The movement of a large group of people from a community usually has dramatic effects on the demographic characteristics of those left behind (Vredenbreght 1964; Lineton 1975a, b; Hugo 1978; Forbes 1981; Naim 1979). The movement also influences the social and economic growth of the place in which the migrants originated. Hugo (1978), among others, studied in detail the impact of migration on villages in West Java. According to him, remittances made up more than 50% of total household income. The bulk of remittances was spent on education or invested in housing. Vredenbreght (1964), Forbes (1981), and Naim (1979) also found that remittances were important features of rural–urban migration. The out-migration of people from the village, mostly young persons, produced problems of selecting village leaders, and problems of raising enough labor for "cooperative work" projects. The rate of divorce also seemed to be increasing in places from where many males have out-migrated (Hugo 1978; Naim 1979).

In the literature, the urban destination has received more attention than the rural destination. Several migration studies in Jakarta (Temple 1975; Papanek 1975; Jones 1978; Oey 1977) showed that in-migration was particularly important in the growth of Jakarta not only in items of population growth but also in urban development. History has shown that Jakarta was built by migrants. Undoubtedly, the large number of migrants led to increasing demand for public facilities such as housing, transportation, and hospitals. Montgomery (1975) showed that Indonesia is experiencing a serious urban unemployment problem among the young: although the evidence was based on rough and satisfactory data. Even though there was a high risk of not getting a job, migration to the cities is still taking place. Another impact of large numbers of poor migrants in the cities, which has not received much attention from researchers, is the sociopolitical impact on national political stability. Jones (1978, 37) showed that in the case of Jakarta and Surabaya, a true urban proletariat divorced from any rural roots is likely to develop. The migrant poor will continue to compete for jobs and services with the local-born poor and also with the urban middle classes. There is a high possibility that this situation can lead to social unrest and political conflicts.

Migration and Increasing Inequality

From the beginning of its five-year development plan in 1967, Indonesia gradually sought to integrate its national development into the regional and global capitalist economy. The substantial lack of capital was one of the reasons for the beginning of the open-door policy to invite foreign loans and investors into the country. Aided by large inflows of concessional finance and substantial foreign private direct investment from Japan, in the 1970s, Indonesia actively exploited its considerable natural resources: the mainstay of which has been oil. The Indonesian economy performed reasonably well during the 1970s, with an average growth rate per capita GDP of 5.4%. This was better than its major ASEAN partners, and approaching those achieved in the NICs.

During the 1980s, the global oil surplus led to a decline in foreign exchange earnings and a resolve on behalf of the government to diversify its export base. Since the early 1980s, a series of reforms have been introduced, which are designed to improve economic efficiency and shift the direction of the economy from import substitution to an export orientation, with a particular emphasis on nonoil exports. Many economic reforms, including deregulation, have been introduced to facilitate operations in the financial sector, promote foreign trade, encourage investment (particularly foreign investment), streamline customs, and shipping, ease regulatory controls on land transport, and improve the country's tax base. From the available evidence, it seems that these reforms have stimulated foreign investment and significantly boosted nonoil exports. By 1995, Indonesia was ranked by the World Bank as a middle-income country with a per capita income of close to US$1000, up from less than US$100 at the time when the New Order began its development planning.[2]

Among economists, however, there have always been different views on the impact of economic reforms on the distribution of income. These divergent views are partly due to the different approaches and methods used to measure the economic impacts on income distribution. Azis, for example, applying the so-called computable general equilibrium (CGE) model, to the Indonesian case, indicates through both the static and dynamic simulations that the postreform progress in the country's macroeconomic condition is likely accompanied by worsening—albeit slightly—household income distribution between income groups. The nonagricultural sector appears to be the major beneficiary of the reform. From the dynamic simulations, a worsening distribution is also found between rural and urban areas. However, the results of both simulations also show that an improved poverty condition is likely achieved following the reform.[3]

[2]Indonesia is labeled by Hill (1996) as Southeast Asia's emerging giant in his recent book, which comprehensively assesses the Indonesian economic achievements since 1966.

[3]See unpublished paper by Iwan J. Azis "Impacts of Economic Reform on Rural-Urban Welfare: A General Equilibrium Framework," Cornell University. No date of publication.

Since the colonial period up to the late 1970s, population mobility in Indonesia can be classified into five major types. First is the population mobility into plantation areas. This movement occurred mainly during the colonial period. The second type is international migration. This movement had a limited impact on contemporary population growth in Indonesia. Thirdly, it is the induced rural to rural movement or the transmigration program. This program is essentially a continuation of an earlier, so-called colonization program established by the Dutch. Under the colonization program, the Dutch had only succeeded in resettling 189,938 families (McNicoll 1968, 62) from Java to the Outer Islands, mainly in Sumatra and Kalimantan. Under the transmigration program, the Indonesia government had only succeeded in resettling around 500,000 families from Java, Bali, and Lombok to Outer Islands. The fourth type of population movement is rural–urban migration, either permanent or circulatory. Lastly, the fifth type of movement is voluntary migration or *merantau* among the highly mobile ethnic groups in Indonesia.

Data on recent migration for the last two decades (Table 4.2) clearly show that the number of migrants for the period of 1985–1990 is larger than the previous two periods (1975–1980 and 1980–1985). The provinces that have a higher rate of recent migrants increase during the 1985–1990 period compared to the two previous periods are Aceh, Riau, Bengkulu, West Java, Central Java, East Java, East Timor, Central and East Kalimantan, South Sulawesi, Maluku, and Irian Jaya. In contrast to other parts of the country, the increasing number of recent in-migration to provinces in the eastern part of Indonesia was already occurring in the 1980–1985 period.[4]

As the government capacity to sponsor population movement through transmigration programs became limited, as oil price collapsed, in the mid-1980s, most recent migrants in the 1985–1990 periods were most likely to be voluntary migrants. The various economic reforms, which have been implemented since the mid-1980s to boost foreign investment, have resulted in the opening up of job opportunities in many places, such as Batam Island in Riau, East Kalimantan, and Irian Jaya. A significant indication that most of the recent in-migration consist of spontaneous or voluntary migrants is their destinations, which are mostly urban areas compared to rural areas. The process of population mobility change from assisted to voluntary migration was accompanied by the process of labor reallocations from low-income agriculture to higher productivity nonagricultural employment. According to Paauw (1963), agriculture is the major source of the labor surplus during a successful transition to modern economic growth. Nonetheless, this sector continued to absorb a large share of labor force growth throughout the 1971–1990 periods. The

[4]In Indonesia, the only source of information on population mobility or migration at the national level is the population census. The population census is taken every 10 years, and the latest was undertaken in 1990. Information on two types of migrants can be derived from the population census. Lifetime migrants are those people whose province of birth is different from the current province of residence. Recent migrants are those people whose place of residence 5 years before the census is different from their current province of residence. In many ways, the data on recent migrants provide more accurate pictures of existing population mobility than the data on lifetime migrants.

Table 4.2 Recent in-migration by province of residence: 1975–1980, 1980–1985, 1985–1990

	In-migration					
	1975–1980		1980–1985		1985–1990	
Provinces	*N*	%	*N*	%	*N*	%
Aceh	51,208	2.2	37,692	0.8	56,326	1.6
North Sumatra	95,586	1.2	59,600	2.6	107,922	1.0
West Sumatra	93,117	3.1	75,757	4.0	129,049	3.2
Riau	98,652	5.2	91,881	11.8	245,465	7.5
Jambi	107,273	8.8	52,647	–	134,397	6.8
South Sumatra	221,165	5.6	105,064	1.9	212,196	3.4
Bengkulu	66,902	10.2	33,386	–	82,831	7.0
Lampung	507,803	13.0	126,677	2.0	212,298	3.5
Jakarta	766,363	13.6	684,001	14.0	833,029	10.1
West Java	551,960	3.2	560,460	3.5	1,350,596	3.8
Central Java	183,761	0.8	171,473	9.5	384,753	1.3
Jogjakarta	98,856	3.8	112,331	2.7	161,740	5.6
East Java	203,175	0.8	165,731	–	328,607	1.0
Bali	37,254	1.7	23,565	1.9	65,967	2.4
NTB	26,221	1.0	26,762	–	37,401	1.1
NTT	25,976	1.0	20,050	19.4	27,107	0.8
East Timor	–	–	13,093	–	26,255	3.5
West Kalimantan	39,380	1.8	19,331	–	43,799	1.4
Central Kalimantan	46,699	6.1	33,328	–	78,791	5.6
South Kalimantan	61,704	3.4	55,752	–	98,330	3.8
East Kalimantan	112,620	10.6	83,976	0.8	194,531	10.4
North Sulawesi	45,498	2.4	14,783	–	34,736	1.4
Central Sulawesi	83,595	7.5	28,067	–	71,143	4.1
South Sulawesi	65,208	1.1	48,453	18.9	119,455	1.7
Southeast Sulawesi	51,014	6.4	69,547	3.9	71,143	5.3
Maluku	46,904	3.8	23,860	2.6	68,701	3.7
Irian Jaya	33,420	3.4	52,771	–	73,776	4.5
Total	3,721,314	2.8	2,790,038	2.0	5,251,195	2.9

Source: Mantra (1992, 55)
Note: The percentage is calculated from the number of populations in each province

1980–1985 periods, especially, was a period of slow release of agricultural labor to nonagriculture. During this period, the agricultural sector absorbed 48% of the labor force increase, 26%, and 29% higher compared to 22% in the 1970s and 19% during the 1985–1990 periods, respectively.

During the 1970s, import–substitution growth was fueled by a booming oil export market. Real gross domestic product in the manufacturing and other nonagricultural sectors grew rapidly compared to agriculture. This situation enabled nonagricultural sectors to support higher growth of employment than agricultural sectors. While agriculture employment grew by only 1.1% per year from 1971 to 1980, employment in nonagriculture grew by an annual average rate of 6.4%. In the 1985–1990 period, labor reallocation from agriculture to nonagriculture regained momentum. Agriculture absorbed only 19% of the new employment, while the nonagricultural sectors, except for trade, achieved substantially higher than its historical contribution. Moreover, manufacturing became the leading source of new employment in the

Table 4.3 Recent in-migration to urban and rural areas by province of residence: 1985–1990

Province	In-migration 1985–1990			
	Urban		Rural	
	N	%	*N*	%
Aceh	22,239	4.1	34,087	1.2
North Sumatra	63,927	1.8	43,955	0.7
West Sumatra	40,472	5.0	88,577	2.8
Riau	97,600	9.3	147,865	6.6
Jambi	35,375	8.2	101,022	6.4
South Sumatra	65,935	3.6	146,261	3.3
Bengkulu	32,956	13.7	49,875	5.3
Lampung	41,755	5.6	170,543	3.2
Jakarta	833,029	10.1	–	–
West Java	1,032,427	8.5	318,169	1.4
Central Java	156,297	2.0	228,456	1.1
Jogjakarta	118,094	9.1	43,646	2.7
East Java	178,256	2.0	150,351	0.6
Bali	48,964	6.7	17,003	0.8
NTB	18,784	3.3	18,617	0.7
NTT	13,550	3.6	13,557	0.5
East Timor	7623	13.1	18,632	2.7
West Kalimantan	20,134	3.1	23,665	0.9
Central Kalimantan	24,132	9.8	54,659	4.7
South Kalimantan	35,828	5.1	62,502	3.3
East Kalimantan	117,491	12.8	77,040	8.0
North Sulawesi	17,095	3.0	17,641	0.9
Central Sulawesi	24,031	8.5	46,003	3.2
South Sulawesi	53,611	3.1	65,844	1.2
Southeast Sulawesi	18,650	8.1	52,493	4.7
Maluku	20,168	5.7	48,533	3.2
Irian Jaya	45,880	11.6	27,896	2.2
Total	3,184,303	5.7	2,066,892	1.7

Source: Mantra (1992, 56)

Table 4.4 Employment absorption by sector 1971–1990 (percent)

	1971–1980	1980–1985	1985–1990
Agriculture	21.8	47.7	19.0
Manufacturing	16.1	10.1	26.7
Trade	19.4	24.8	14.6
Services	25.0	10.1	16.2
Others	17.7	7.3	23.5

Source: Paauw (1963, 316)

nonagricultural sector for the first time. This dramatic increase in manufacturing employment was accompanied by high rates of employment creation in construction (8.3% of the total) and in the transport/communications sector (7.8% of the total). These recent improvements in nonagricultural employment creation were the result of a drastic change in the economy's growth regime and the change from import substitution to export promotion (Tables 4.1 and 4.4).

Population mobility and employment data as revealed from population censuses and labor force surveys during the last two decades clearly indicated that the mode of population mobility was strongly associated with the broader socioeconomic processes instigated by government policies.[5] The increase in the volume of population mobility, as mentioned earlier, is essentially the logical consequence of three factors: first, the surplus of labor; second, the improvement of transportation means and network; and third, the opening up of economic activities, particularly in urban areas where the informal sector economy provides job access elasticity for migrants.[6]

Conclusion

In Indonesia, the interrelationship between migration and economic development is not always clear. Population mobility can play a major role in economic development, for instance, in the case of colonization and transmigration, many migrants in the place destination resulted in a new agricultural area was developed. On the other hand, the development of a new plantation area or industrial area has attracted many migrants from the surrounding areas to provide labor. So that development projects are sometimes a factor encouraging migration, while migration may also lead to the establishment of development projects. The relationship between migration and development becomes more complex when the society is developing, for example, when education, communication, and transportation become "modern." It is also seen that Indonesia, like many other developing countries, is experiencing the effect of urbanization. In the context of population mobility, perhaps, urbanization (mainly as a consequence of rural–urban migration) is the most important problem. Instead of transmigration, urbanization seems to be the popular field among the researchers.

In this review, transmigration has been given only a little attention because transmigration is much more of an involuntary than voluntary movement. In addition, transmigration has only a small effect in reducing the population pressure in Java, as well as on the whole process of development. However, in literature, transmigration has always been given a lot of attention by some researchers.[7] In the case of voluntary migration among the ethnic groups (*merantau*), the cause of out-migration should be seen with caution. The traditional factors, such as the social system and the social norms, were not the only factors affecting their migration

[5] The preliminary result of the 1995 intercensual survey shows the continuation of migration trends and patterns between 1990 and 1995. (See unpublished discussion paper, "Inter-provincial Migration in Indonesia: Trend and Pattern." 1995. Central Bureau of Statistic, January 1997.)

[6] Since the 1980s, large numbers of migrant workers have moved across the national boundaries, particularly to Malaysia. A government source estimated that the number of Indonesian migrants in Malaysia is around 1.5 million, of whom 600,000 are illegal (*Kompas*, March 15, 1997). The social and political impacts of the influx of Indonesian migrants in Malaysia are significant due to the crucial ethnic balance between the Malays and the Chinese (see Chap. 8 on the politics of Indonesian illegal migration to Malaysia).

[7] There are at least two annotated bibliographies on transmigration alone; see Meyer and MacAndrews (1978) and Budiman (1985).

behavior. The impact of economic development and the sociopolitical change from the wider society must be considered in viewing the dynamics of population mobility. Finally, the role of government policy, particularly since the New Order, such as the provision of development projects in Java or outside Java and rural or urban areas, needs to be assessed in some more detail to facilitate the understanding of population mobility in Indonesia.

Chapter 5
Underdevelopment and the Politics of Migration in Eastern Indonesia

Eastern Indonesia is a region of distinct natural beauty, heterogeneous population, and diverse culture. Since the early 1990s, the contrast between Eastern and Western Indonesia seemingly replaced the classic dichotomy of "inner" and "outer" Indonesia as the central government began to pay attention to the backwardness of the eastern provinces. Under the New Order, development was reduced to merely economic growth, and development progress was therefore measured narrowly by macro-social–economic indicators. The spatial dimensions and the regional diversity were neglected, and this has resulted in the widening development gap between the western and the eastern provinces. Although the central government began to pay attention to the development problems in the eastern provinces, real policy implementation, however, is still strongly constrained by the existing centralized and narrow-minded development paradigm.

In the last three decades, the number of people who migrated between provinces in Indonesia has shown a constant increase. Since the early 1980s, apart from Java, which has always been a major migration destination, Eastern Indonesian provinces have also become an important interprovincial migration destination. Although after the mid-1980s transmigration programs were scaled down, spontaneous migration to Eastern Indonesia continues to occur. Two decades ago, Titus (1978), in his findings based on an analysis of the 1971 population census, argued that interregional migration in Indonesia reflects the degree of social and economic development between regions. According to Forbes (1981), Titus' findings significantly demonstrate the need to examine the process of uneven development in Indonesia: particularly the forces that have given rise to the process and that seem to be perpetuating it. Major studies on migration in Indonesia, however, tended to overlook the social and political dimensions, and emphasized more the demographic and economic aspects of population mobility. The domination of development economic

R. Tirtosudarmo, *From Colonization to Nation-State*,
https://doi.org/10.1007/978-981-16-6437-3_5

frameworks during the New Order era apparently influenced migration study, which mostly focused on the demographic and economic aspects.[1]

Eastern Indonesia became a topic of wide public discussion among academics and development specialists after President Suharto announced he would provide more attention to the development of the region. Although many policies have attempted to bolster economic development in Eastern Indonesia, their implementation generally is still unsatisfactory. There is no doubt that the New Order's official statistics clearly indicate the increase in people's welfare in Eastern Indonesian provinces. However, as many observers have noted the social and economic inequality between regions and groups of the population is also increasing. The various social and political conflicts that have intensified in several places in the last 5 years are strong indications that genuine sustainable development is not taking place in Indonesia. The ideology of development adopted by the Indonesian state in the last three decades has proved to be a failure as clearly demonstrated by the current political and economic crisis that culminated in President Suharto's abrupt resignation on May 21, 1998.

The Underdeveloped Region

In January 1990, at a parliamentary meeting, the president officially admitted that Eastern Indonesia is being an economically left behind region. Comparing human resource indicators for Eastern Indonesia with Java-Bali, Sumatra, Kalimantan, and Indonesia as a whole, it clearly shows that Eastern Indonesia has slower development in many respects.[2] In terms of annual population growth between 1971 and 1992, except for North and South Sulawesi and East Nusa Tenggara, which have similar rates of growth to Java-Bali, all other provinces in Eastern Indonesia have similar rates of growth to Sumatra and Kalimantan, which are above the national level.

The rates of annual population growth during the 1971–1992 period indicate a diversity of population dynamics in this area. Assuming that the natural increase is evenly distributed, therefore, migration plays the most important role in affecting the dynamics of the population. The high annual population growth in provinces such as

[1] Migration analysis in Indonesia is generally constrained by the limited sources of data. The only available source of information on population mobility at the national level is population censuses. The migration data collected in the population census are limited only to the permanent movement between provinces. Information on nonpermanent movement or circular migration and district-level administration is not available from population censuses. The relevant questions applicable to migration analysis are associated with birthplace, place of current residence, place of previous residence, and place of residence 5 years ago.

[2] In this paper, the Wallace line, which lies between Kalimantan and Sulawesi, is used to divide Indonesia into Western and Eastern parts. This is slightly different from the government's definition, which included Kalimantan as part of Eastern Indonesia.

Central and Southeast Sulawesi as well as Maluku could be attributed to the high rate of in-migration, particularly through the transmigration scheme, into these provinces. On the other hand, a substantial flow of out-migration is likely the reason for the low rate of annual population growth in provinces such as South Sulawesi, North Sulawesi, and East Nusa Tenggara.

In terms of per capita income, except for Irian Jaya, all provinces in Eastern Indonesia are below Java-Bali, Sumatra, Kalimantan, and national average (Table 5.1). While all provinces in Sulawesi can be classified as being in the middle-income group, West Nusa Tenggara, East Nusa Tenggara, and East Timor are the poorest provinces in Indonesia. However, from the perspective of human resources, income per capita can be a misleading indicator as it does not represent the actual welfare condition of the population. Eastern Indonesia such as Irian Jaya and Maluku are good examples. The high income per capita in Irian Jaya and Maluku, even after oil revenue has been excluded, is likely still to be heavily influenced by the revenue from copper and gold mining (Freeport McMoran Inc., in Irian Jaya), timber exploitation, and fisheries. With their relatively small populations, it is not a surprise for Irian Jaya and Maluku to have a high per capita income. In the case of West Nusa Tenggara, East Nusa Tenggara, and East Timor, the low per capita incomes perhaps genuinely reflect their human resource development, except for some parts of Lombok. These three provinces are similarly affected by their poor natural resources. Poverty and economic hardship in West and East Nusa Tenggara have become a strong push factor for people to migrate to other places to seek employment. The people who live on the small island to the east of Flores are well known for their tradition of migrating to Johor in West Malaysia and Sabah in East Malaysia (Tables 5.1 and 5.2).

Except for major urban areas, in most parts of Eastern Indonesia where the economic situation of the population is relatively poor, the education facilities and the availability of transportation are also generally poor. The problem is less felt, however, in areas close to urban centers. In the provincial capital cities, the population's access to all levels of education is not a problem. Tertiary education is accessible in all provincial capital cities, as well as in several district capitals. The problem of accessibility for most senior high school graduates entering tertiary education in their province is a result of the limited number of places available and the lack of transportation for people who are living on different islands or in remote areas. For some high school graduates who come from better-off families, entering tertiary education outside their provinces often becomes their ambition. They perceive that universities, particularly in Java, are more prestigious compared to universities in their provinces.

Besides education, human resource conditions are reflected by the level of the infant mortality rate and the availability of health facilities, such as community health centers, hospitals, and medical doctors. In the national contexts, North Sulawesi and South Sulawesi are two provinces in Eastern Indonesia that have a relatively similar rate to Java-Bali, Sumatra, and slightly lower than the national average level. The number and availability of health centers and other medical facilities in Eastern Indonesia are also slightly lower than in Java-Bali and Indonesia

Table 5.1 Eastern Indonesia: demographic and economic indicators, 1992

Province/region	Total population ('000)	Population annual growth 1971–1992	Population density (persons/km²)	GRP per apita (Rp. 1.000 constant 1983 price)		Labor force participation rate	Daily minimum wage (Rp) 1994
					Incl. oil		
North Sulawesi	2550	1.85	134	449	–	53.1	–
Central Sulawesi	1803	3.22	25	381	–	56.3	2.300
South Sulawesi	7229	1.55	99	458	–	48.5	2.300
Southeast Sulawesi	1450	3.32	52	429	–	58.2	2.800
Maluku	1954	2.73	26	506	511	53.1	3.100
Irian Jaya	1765	–	4	699	794	61.3	–
West Nusa Tenggara	3493	2.17	173	274	–	62.5	2.350
East Nusa Tenggara	3403	1.83	71	243	–	66.1	–
East Timor	789	–	53	220	–	65.5	3.400
Jawa-Bali	110,720	1.73	824	622	657	58.5	3.275
Sumatra	38,366	2.88	81	538	914	57.6	2.958
Kalimantan	9684	2.95	18	728	1.128	59.9	3.083
Indonesia	186,042	2.11	97	588	711	57.3	2.945

Source: Tirtosudarmo (1995, 7)

Table 5.2 Eastern Indonesia: education and health indicators, 1992

Province/region	Education enrollment per 1000 persons			IMR	Health Center '000 pop	Hospital (and hospital beds)	Number of doctors (1991)
	Primary	IMR	Senior High				
North Sulawesi	141	37	21	55	0.051	22 (2.625)	373
Central Sulawesi	168	31	12	77	0.046	16 (1.131)	191
South Sulawesi	147	33	18	58	0.038	56 (4.884)	651
Southeast Sulawesi	166	35	17	64	0.061	12 (2.678)	105
Maluku	182	40	20	61	0.060	18 (1.543)	144
Irian Jaya	145	33	14	72	0.084	21 (1.603)	138
West Nusa Tenggara	145	21	12	114	0.029	12 (794)	177
East Nusa Tenggara	171	26	10	61	0.046	23 (1.664)	156
East Timor	123	38	4	67	0.077	10 (549)	106
Jawa-Bali	129	28	13	49	0.026	463 (65.186)	10.591
Sumatra	164	34	16	58	0.034	255 (24.577)	3.148
Kalimantan	158	27	13	60	0.061	77 (6.226)	858
Indonesia	141	29	14	60	0.032	985 (11.146)	

Source: Tirtosudarmo (1995, 8)

generally. In the regional context, North Sulawesi and South Sulawesi show the lowest infant mortality rate, also coinciding with the best health facilities. In contrast, West Nusa Tenggara shows an extremely high rate of infant mortality, implying the worst physical human resources. In fact, the number of health facilities available in West Nusa Tenggara is not too critical when compared with provinces such as East Timor or Southeast Sulawesi. However, the infant mortality rate in West Nusa Tenggara is remarkably high, almost double the rate for the national level.[3]

The statistical indicators clearly indicate a significant improvement in education and health conditions, and genuine development in human resources in Eastern Indonesia, however, is still a problem. The strategy adopted by the New Order government has been based on the assumption that economic growth should be the prime mover of national development. The emphasis on economic growth has resulted in the high priority given to the improvement of infrastructure, both at the national and regional levels. This, in turn, has resulted in the lower priority given to human resource development. The presidential statement on Eastern Indonesia's development has successfully attracted a wide academic debate; however, real moves by the government to accelerate the pace of development in eastern provinces

[3]The level of outside interventions, from both the government and nongovernment organizations to improve health conditions in West Nusa Tenggara is, in fact, quite remarkable. Therefore, it is no wonder that West Nusa Tenggara has become of interest to health development observers. West Nusa Tenggara represents a good example of the complicated and multifaceted linkages between internal cultural values and the various external interventions to increase the social well-being of the population. It seems that a great deal of outside intervention is not a guarantee for the improvement of human resource conditions. In such a situation, the attitude of the officials, people's habits, and social norms have their dynamics, which can reverse the outcomes of the intervention.

have surprisingly been terribly slow to emerge. Part of the problem is seemingly the incapacity of the central government to come to grips with the task, given the social and geographic complexity of what is officially designated as the Eastern Indonesia regions (*Kawasan Timur Indonesia*). The area consists of not only resource-rich provinces, such as East Kalimantan and Irian Jaya, but also resource-poor provinces, such as East Nusa Tenggara and East Timor. Furthermore, the centralist character of the national government, which tends to standardize regional policy, in which rich and poor provinces are treated uniformly, might have generated economic inefficiencies.

Eastern Indonesia Within New Order's Central–Regional Contexts

At the declaration of independence on August 17, 1945, Indonesia's founding fathers stated that Indonesia was to be a unitary state. Yet, the struggle to materialize a genuinely integrated nation-state proved to be a long and complicated political process. For example, the war against the Dutch continued until 1949, when de facto independence was finally achieved. The period 1950–1957 was one of parliamentary democracy, which was marked by constant political conflict both among the political parties and between central and provincial governments. There are two major regional conflicts occurred during this time: the PRRI/PERMESTA in West Sumatra, South and North Sulawesi, and DI/TII in West Java. The guided democracy period of 1957–1965 that followed firmly established the strong central government *vis-a-vis* the weak regional government. The outcome of the political crisis in 1965 that toppled Sukarno and at the same time elevated General Suharto as the president further established the domination of the state vis-a-vis the civil society. Since 1967, under Suharto's presidency, the political system called Pancasila democracy was engineered in such a way as to establish him as the sole authority in the country. During the New Order, separatist movements were raging in Irian Jaya and East Timor (Eastern Indonesia), and Aceh (Western Indonesia). These separatist political movements, however, did not have a significant effect on challenging the national government.

According to Mackie (1980), since about 1970, several important development changes in Indonesia have greatly strengthened the central government's capacity to control the regions. First, the tightening of central authority within the military has significantly reduced the power of regional commanders over their territories. Second, the process of depoliticization has eliminated the possibility of a political party being able to stir up mass grievances against the central government. Third, the improvement in interisland shipping and airline services, as well as telecommunications, commercial, and banking networks, has played an important role in integrating the regions into a single economy. In addition to three important centrifugal factors indicated by Mackie, the use of *Bahasa Indonesia* as a national language has played an integrating role as a tool of communication among peoples of such a plural ethnolinguistic society.

The national political system, which is highly centralistic and tightly controlled by Suharto and the military, has directly influenced the structure of central and provincial-level government relationships. Furthermore, such an unequal relationship also affects regional development planning formulation and implementation. In 1974, the government promulgated Law No. 5, which designated the legal framework for the role and authority of the provincial governments. According to this law, the provincial government authorities exercise only those powers that have been granted to them by the central government. The national government, primarily represented by the Department of Home Affairs, acts through the governors and has the responsibility to supervise and to support all development activities carried out in the provinces. Development activities at the provincial level, based on Law No. 5, 1974, are classified into three broad categories. The first category is the development activities and programs, which are implemented by the sectoral line agencies in the regions. In this category, the central government is responsible for all related development activities, such as planning, implementation, control, and financial supports. The second category is the development activities that are funded and implemented by the district-level government technical offices (*Dinas).* The third category is the development activities that are supposed to be controlled by the provincial governments but which, because of their limited financial capacity, are handed over to the central government to control. An example of such activities is the so-called *Inpres* (presidential instruction) Project, which is implemented by provincial governments under the control and supervision of the Ministry of Home Affairs.

In 1976, the central government created Regional Development Planning Boards (*Bappeda*) to strengthen the provincial government's abilities in planning and coordination of development programs. In theory, the Regional Development Planning Board should perform several functions, such as developing basic regional development plans, coordinating planning among the sectoral line agencies at the provincial levels, drafting the regional development budgets, and monitoring and evaluating development. In practice, however, what looks neatly described as bottom-up planning does not really take place. The actual planning is still largely a top-down process. Development programs and projects at the provincial level are mostly designed and planned by the National Planning Board (*Bappenas*) and various ministerial offices in Jakarta.[4]

The central government procedure in allocating provincial budget is also criticized and it did not consider the potential ability of the regions. Pangestu and Aziz (1994), for example, argue that the central government's planners are not considering incentives for local resource mobilization as each region receives the same

[4]Many sources note that the personal lobbying of the provincial bureaucratic leaders, particularly the governors, is especially important in determining project and budget allocations for their respective provinces.

amount per head of population, especially in the case of package development grants (*Inpres* Project). The primacy of population size as the major criteria in allocating provincial budgets, therefore, not surprisingly, ensured that the largest proportion of the budget still poured into Java. Furthermore, this imbalance is reinforced by the lack of an incentive criterion, such as tax efforts, and the failure to consider the regional potential in fund allocations. Thus, the financial transfer from the center to the regions plays no significant role in reducing the existing regional disparity, particularly between Western Indonesia and Eastern Indonesia.

The president, in a parliamentary meeting in January 1990, formally stated the government's recognition that Eastern Indonesia's provinces had been economically left behind by the western provinces. In his speech, the president argued that the pace of development in Eastern Indonesian provinces had to be accelerated in order to narrow the development gap between eastern and western provinces. The policy to develop Eastern Indonesia was formally stated in the Broad Outlines of State Policy (*GBHN*) as the basis for the sixth five-year development plan (1994–1988). It is stated explicitly in the sixth plan that priority should be given to the expansion of transport systems in Eastern Indonesia. An explicit policy statement on transmigration, for example, clearly shows the trend toward Eastern Indonesia: *"The policy to direct transmigration to Eastern Indonesia is intended to develop a better economic and socio-cultural infrastructure; to accelerate communication and transportation; to develop social institutions, such as co-operation, financial institutions, and market facilities to enhance the bargaining power to the society."*[5]

After the 1990 presidential speech, a great deal of discussion on development in Eastern Indonesia flourished among economists and development experts. Although the presidential statement on Eastern Indonesia's development successfully attracted wide academic debate, real moves have surprisingly been slow to emerge. The slowness of the decision-making process in the central government, for example, is indicated by the creation of the National Advisory Council on Development of Eastern Indonesia (*Dewan Nasional Pembangunan KTI*) in 1993. The council, which was headed by President Suharto, consists of no less than 12 ministers and clearly reflects the consensual nature of the policy on Eastern Indonesia. In addition, the appointment of Professor B.J. Habibie—the State Minister for Research and Technology—originally from Eastern Indonesia, and three former governors from Eastern Indonesian provinces as advisors, shows the symbolic approach of the central government in dealing with development in Eastern Indonesia. Furthermore, the central government decision on the provinces designated as Eastern Indonesia, which eventually also included all the provinces on Kalimantan Island, raised the question of the effectiveness of the policy.[6] The nature of both the organizational

[5]The sixth five-year development plan, Book I, p. 55.

[6]In a seminar on Eastern Indonesia's development at LIPI in 1996, former governor of South Sulawesi, and advisor of B.J. Habibie, Amirrudin, acknowledged that the reasons to incorporate Kalimantan into Eastern Indonesia were merely because the bureaucratic elite in Kalimantan provinces protested of being neglected by the central government if not included into the Eastern Indonesia development plan.

structure and the wide area designated as Eastern Indonesia clearly reflects the usual consensual practices adopted by the central government to solve the classic problem of coordination between departments, which is crucial to tackle the complex task of uplifting development in Eastern Indonesia.[7]

At a parliamentary session on August 16, 1990, Suharto stated the intention to give more autonomy to the regions. The regions concerned here are the second-level regional governments (district), which according to the National Law No. 5, 1974, are considered autonomous. Yet as many observers have noted that the intention to give greater opportunity to the regions by the Suharto government was merely a political gimmick with no substantial effect on existing unequal central–regional government relationships. The second-level regional government, though legally stated as autonomous, in practice is economically dependent on both the provincial and central governments.[8] The implications of the New Order's bureaucratic arrangement, particularly on the suppression of the local people's political participation, are obvious. As shown by the findings of a study by the Center for Regional and Political Studies, Indonesian Institute of Sciences, the political sphere of society at the regional levels is very limited and the local people are therefore further marginalized.[9] A recent study by another research team of the Indonesian Institute of Sciences indicated that the central government's regional autonomy project was mainly dealt with the technical administrative aspects of the regional autonomy and clearly will not have a significant effect on the existing central–regional relationships.[10] The unequal relationship between the state and the civil society, as well as between the center and the regions, however, has been strongly questioned following the collapse of the Suharto regime in May 1998. A new political equation is apparently on the horizon, and its implication for the future path of regional development is therefore becoming inevitable.

Population Mobility in Eastern Indonesia

Based on the result of the 1971 population census, Titus (1978), using Amin's center–periphery approach (1974), has tried to show that development strategy and foreign economic activities play an important role in affecting interregional population movement in Indonesia. The most favored foreign company investments are in the extractive industries, such as timber and oil, as well as the assembling and

---

[7]It is then no surprising that such a high-level organization will face problems in its decision-making process. For example, it took more than 1 year after its designation for this organization to hold its first plenary meeting.

[8]A study conducted by the Center for Regional and Political Studies, Indonesian Institute of Sciences, in 1992, shows that the district-level government cannot be called an autonomous region as it has only having a quasi-autonomous governmental control (Sulistyo 1995).

[9]See Sulistyo (1995).

[10]See PEP-LIPI (1997).

construction industries, followed by commerce and services. These activities have attracted a great deal of foreign capital. The development of extractive industries is almost exclusively limited to the Outer Islands, while the other activities being mainly in the urban centers of Java. Titus classified a province as center or periphery based on the presence of such economic activities. In addition, every region was assessed on the existence of "attraction factors" (i.e., employment opportunities, regional per capita income, urban attraction) or "expulsion factors" (i.e., rural population pressure, ecosystem instability, lack of education facilities, political upheavals).

According to Titus, the greatest mobility together with net in-migration was to be found in the economic "boom" provinces of both the center type (Jakarta, North Sumatra) and the relatively developed periphery type (South Sumatra, Riau, East Kalimantan). The lowest mobility and a zero-migration balance were to be found in the isolated and still largely self-sufficient periphery type of provinces (i.e., East and West Nusa Tenggara). The highest mobility together with net out-migration appeared in the highly integrated but stagnating peripheral provinces close to center groups (i.e., West Sumatra, Central Java, and Jogjakarta). Criticizing the work of Titus, which had ignored short-term and nonpermanent movements, Forbes (1981), using the same approach, conducted a study in South Sulawesi, mainly in Ujung Pandang. He focused attention on the circulation of petty traders between Ujung Pandang and the hinterland and showed the majority of migrant, petty producers kept in contact with their villages with regular trips home. There was important circulation between Ujung Pandang and the hinterland not only of labor but also of cash and goods. The main theme of Forbes' study is the importance of viewing population mobility as a reflection of the structure and processes of Indonesian society.

The general pattern of migration trends between Java and other islands shows an interesting shift after the mid-1980s. The net out-migration from Java to other islands, between the periods 1970–1980 and 1980–1990, markedly declined from 137 to only 15.7%.[11] This change reflects the increasing pull factors in Java and the decreasing role of the transmigration policy as the major push factor for out-migration from Java, particularly after the mid-1980s. Eastern Indonesia as part of the Outer Islands also shows a similar pattern and trends. In more recent developments, based on the 1995 Intercensal Population Survey (*SUPAS*), Raharto and Romdiati (1998) show that in-migration to Eastern Indonesia's provinces was also significantly declining between 1990 and 1995. In 1990, in-migration to Eastern Indonesia accounted for 54%, whereas out-migration to Western Indonesia was only 36%. In 1995, in-migration to Eastern Indonesia dropped to 49%, while out-migration significantly increased to 49%.[12] The changes in the volume and pattern of migration between Eastern Indonesia and Western Indonesia strongly

[11] a detailed analysis of the changing patterns and processes of migration in Indonesia during the New Order see Hugo (1997).

[12] See Raharto and Romdiati (1998).

indicate that the regional disparity between Eastern Indonesia and Western Indonesia is increasing. The general pattern of the migration phenomena apparently reflects that the New Order's regional development has basically failed to achieve more equally distributed economic benefits between Java and the Outer Islands, as well as between Western Indonesia and Eastern Indonesia. The shift in macroeconomic policies from *import substitution* into *export orientation* policies, after the collapse of the world oil prices in the mid-1980s, has been among the major factors that resulted in the further concentration of manufacturing industries in Java rather than decentralizing the regional economic growth centers in the Outer Islands.

In terms of population size, according to the 1995 Intercensal Population Survey, Eastern Indonesia comprised only 13.3% of Indonesia's total population. From a national demographic perspective, Eastern Indonesia is comprised of nine provinces located in Sulawesi, Maluku Islands, Irian Jaya (formerly West Papua), and Nusa Tenggara islands (excluding Bali); therefore, it cannot be simply put into one distinctive demographic system. For example, the population density varies: Maluku and Irian Jaya together show the least densely populated areas in Indonesia.[13] The Javanese dominated bureaucrats in Jakarta, and therefore, they easily perceive Eastern Indonesia as empty land. Such a perception was apparently behind the development planning decision in the 1980s to redirect the transmigration policy to the Eastern Indonesia provinces.[14]

The impact of the central government's decision to bolster the transmigration program in Eastern Indonesia is clearly shown by the positive net migration trends since 1980 in designated destination provinces, such as Irian Jaya, Maluku, Southeast, and Central Sulawesi. In-migrations apparently also play an important role in the relatively high annual population rate of growth in those provinces. On the other hand, in provinces like North and South Sulawesi, as well as in West and East Nusa Tenggara, negative net migration has also contributed to the low level of annual population growth in those provinces.[15] As a result of the collapse of world oil prices in the mid-1980s, the national budget for the transmigration program was drastically cut back, which, in turn, substantially reduced the number of transmigrant families that had originally been planned for resettlement.[16] The decision to move

[13]In 1995, the density of both Maluku Islands and Irian Jaya was only eight people per square kilometer, while Java was 850 and Indonesia was 95.

[14]The surge in the transmigration target to Eastern Indonesia in the early 1980s was supported by the fact that the central government was noticeably confident of its financial ability, as the World Bank fully committed to back up the program. The government also declared that traditional transmigration destination areas such as Lampung and South Kalimantan were closed for new transmigrants.

[15]East Nusa Tenggara, and more recently followed by West Nusa Tenggara, is becoming the major sending areas of overseas migrant workers to Malaysia and Saudi Arabia.

[16]The polemics on the place and roles of the transmigration program in the discourse of regional development in Indonesia briefly surfaced among Australian academics, in the *Bulletin of Indonesian Economic Studies* in the late 1970s and early 1980s. The views debated in this polemic, unfortunately, were mainly confined to the realms of demographic and economic issues, undermining the critical political aspects of the program.

transmigrant to Eastern Indonesia significantly contributed to the increased transmigration budget as the transportation cost has become very expensive.

There is a strong indication that the decision to resettle transmigrants to Irian Jaya was motivated for security and military reasons. The perceived increasing separatist movement under the Free Papua Movement (OPM) in several places in Irian Jaya obviously worries the military elite in Jakarta as it could provoke a larger political stability in Irian Jaya.[17] The selection of transmigration sites along the border with PNG clearly indicates military intervention in the planning of resettlement areas. The important role of transmigration for national defense and security was explicitly stated in 1985 by the Chief of Armed Forces, General Moerdani, who said that transmigration was the only policy within the economic development framework, which had a direct linkage with national security and defense. Moerdani argued that the military needed to be involved in site selection primarily because transmigration location had a strong relation with the concept of territorial management (Table 5.3).[18]

In the view of the military, geographic integration, in which transmigration policy plays an important role by providing human resources, is a necessary condition for the attainment of national integration.[19] Another important element of this military thinking is the concept of "basic defense." The basic defense of the Indonesian armed forces, according to Crouch (1986), was essentially based on the doctrine of territorial warfare. This doctrine required that the defense forces, especially the army, be organized principally along territorial rather than functional lines. The whole nation, under this doctrine, was divided into territorial commands parallel to the civilian administration. In terms of the defense doctrine, a major function of the territorial commands was to maintain contact with the local people so that they could mobilize quickly to support guerrilla operations whenever necessary. It may be added, of course, that this territorial structure has also been very convenient for political control. The role of transmigration in the concept of basic defense is to populate the empty regions to support the territorial military commands.[20]

In the case of Irian Jaya, the transmigration policy by the central government has contributed to the increase in ethnic tensions. In the mid-1980s, as Irian Jaya became

[17] In the early 1980s, OPM was highly active in the border areas. The military conducted harsh operations to curtail OPM insurgencies. In 1984, thousands of Irianese were reported to have taken refuge in PNG. On the Irian Jaya-PNG border's issues, see studies by PPT-LIPI (Center for Population and Manpower Studies), 1997.

[18] Moerdani's statement was part of his keynote address at the National Defense Institute's seminar (*Kompas*, March 8, 1985).

[19] For the explanation on the conceptual links between transmigration policy and national integration, see Sugardo (1986).

[20] Tanter (1991, quoted from Sebastian 1996, 88–89) describes three elements that influence the territorial strategic concept of the military in Indonesia. According to Tanter, *geography* as the framework, within which *demography* as the content, and *social conditions (IPOLEKSOSBUDMIL)* as the factor of social life that results from the synthesis of the other two elements.

Table 5.3 Religious affiliation by provinces, 1991[a]

Provincial-level unit	Muslim	Protestant	Catholic	Hindu	Buddhist	Other	Total
Aceh	97.75	1.50	0.25	0.02	0.47	0.01	100.0
Bali	5.22	0.58	0.47	93.18	0.55	–	100.0
Bengkulu	97.38	1.32	0.62	0.22	0.46	–	100.0
Irian Jaya	14.90	64.30	20.70	0.14	0.10	–	100.0
Jakarta	84.83	5.94	4.76	0.96	3.51	–	100.0
Jambi	92.16	0.73	0.73	0.20	0.34	5.84[b]	100.0
West Java	97.70	1.10	0.50	0.11	0.50	0.20	100.0
Central Java	95.50	2.20	1.60	0.20	0.40	0.10	100.0
East Java	96.70	2.00	0.70	0.40	0.20	–	100.0
West Kalimantan	54.02	9.00	19.29	0.12	2.70	14.87[b]	100.0
South Kalimantan	97.61	0.82	0.34	0.30	0.20	0.73[b]	100.0
Central Kalimantan	67.70	15.40	1.00	15.80	0.10	–	100.0
East Kalimantan	85.68	9.54	4.01	0.22	0.55	–	100.0
Lampung	94.40	1.60	1.00	2.37	0.60	0.10	100.0
Maluku	54.80	40.90	4.00	0.10	0.10	0.1	100.0
West Nusa Tenggara	95.90	0.60	0.40	2.90	0.10	0.1	100.0
East Nusa Tenggara	9.80	27.60	54.20	0.11	0.15	8.4	100.0
Riau	87.30	2.30	0.50	1.31	7.30	2.70	100.0
South Sulawesi	88.40	8.50	1.40	0.50	0.30	1.00	100.0
Central Sulawesi	76.00	20.30	0.30	2.80	0.20	0.4	100.0
Southeast Sulawesi	98.00	1.30	0.30	0.40	0.05	–	100.0
North Sulawesi	44.10	49.10	2.90	0.58	0.20	0.1	100.0
West Sumatra	97.90	0.90	0.90	0.20	0.30	–	100.0
South Sumatra	94.30	1.30	1.10	0.30	1.80	1.30	100.0
North Sumatra	63.22	27.96	4.60	0.41	3.68	0.13	100.0
East Timor	1.70	2.60	91.40	0.30	0.10	3.90[b]	100.0
Jogjakarta	91.20	3.50	5.00	0.30	0.10	–	100.0

Source: Frederick and Worden (1993, 353), based on information from Indonesia. Department of Information, Directorate of Foreign Information Services, *Indonesia 1992: an Official Handbook*, Jakarta (1992, 55–72)

[a] Figure may not add to totals because of rounding. The source the information used in this table did not provide totals. The last year for which total figures on religious affiliation were made available by the Indonesia government is 1985

[b] No "other" category was reported. The figure represents the estimated difference between the total of the other religions and 100% when data provided by the source did not add to 100

the main destination of the transmigration policy, strong resentment from the local population was aroused. The resentment of the local population to the transmigrants, who mostly originated from Java, was partly due to the local people's perception that they were being unfairly treated by the government. The local population perceived that the transmigrants received more help than the locals. Controversial issues, such as "Javanization," "internal colonization," and "Islamization," were strongly voiced toward the Indonesian government by many foreign NGOs. The World Bank, the main financial supporter of the transmigration program, was also criticized by both foreign and Indonesian NGOs as supporting the destruction of indigenous peoples and tropical forests in Indonesia.[21]

Even though transmigration achievements have been scaled down since the mid-1980s, the net in-migration to Eastern Indonesia is positive, particularly to urban destination areas. As shown by Raharto and Romdiati (1998) in their analysis on interprovincial migration in Eastern Indonesia, the shift of migration to urban areas was strongly identified in provinces such as Irian Jaya, Maluku, and East Timor. The shift from rural to urban areas, because of the decline of the transmigration program, also reflects the elasticity of the urban economy, most notably in the informal sectors, in Eastern Indonesia. The labor surplus, particularly as the proportion of the new working-age cohorts is increasing in Western Indonesia and Java in particular, could be a significant determinant for the young to search for economic opportunities by migration to Eastern Indonesia. The bulk of the migrants to urban areas are likely engaged in many kinds of informal sector activities. In this regard, migrants tend to be more enterprising than the locals due to previous experience in their place of origin. The migrants generally also have higher educational attainment compared to the locals.[22]

Around 30% of the population of Eastern Indonesia resides in South Sulawesi. Given its largest population and its central location, South Sulawesi is therefore particularly important within the *migration system* operating in Eastern Indonesia. The central location of South Sulawesi, of which the harbor of Ujung Pandang is the third largest seaport after Jakarta and Surabaya, and the prominent migrating behavior of its people, provides this province with a dominant role in the development of Eastern Indonesian provinces, both economically and politically.[23] South Sulawesi is also the home of three ethnic groups that are well known for their migrating behavior, namely the Bugis, the Butonese, and the Makassarese. In Eastern Indonesia, these three ethnic groups are popularly labeled as the *BBM,* an acronym of *Bugis–Buton–Makassar*. The notorious *BBM* migration behavior and its social and economic implications in Eastern Indonesia are very well depicted in various studies. Aditjondro, for example, provides a lively description of how the *BBM* migrants conquered socially and economically local peoples in Irian Jaya.[24] In Maluku, and most notably in Ambon, the Butonese migrants comprise the major population group who engage in unskilled jobs and the informal sector economy.[25] After the annexation of East Timor to Indonesia in 1976, the *BBM* migrants apparently were the group most prepared to enter into the new territory to fill the business sectors

[21]For a useful reference on the regional development in Irian Jaya under the New Order, see Manning and Rumbiak (1989).

[22]See Raharto and Romdiati (1998).

[23]On the role of Ujung Pandang as the center of the maritime industry in Indonesia, see Dick's articles on prahu shipping in Eastern Indonesia (1975).

[24]Aditjondro's monograph is a useful source on voluntary migration of the Buginese, the Makassarese, and the Butonese in Eastern Indonesia, particularly in Irian Jaya (see, Aditjondro 1986).

[25]For a detailed population composition in Ambon, see Djohan (1994).

presumably left by the Chinese Portuguese after Indonesia occupied the territory. In East Timor, the migrants mostly originated from Java, Bali, and East Nusa Tenggara, while transmigrants, apparently due to social and political reasons, were mostly recruited from Bali. Up to 1994, there were 1634 transmigrants from Bali who are Hindu, and 1212 transmigrants from Java with most of them are Catholic. The government also relocates the local population under the so-called local transmigration scheme.[26]

The concentration of migrants in Irian Jaya occurred, among others, in Timika where the copper mining industry by an American company—Freeport Inc.—is located. Since the 1980s, the population growth in Timika has increased dramatically in conjunction with the increase in economic activities in the area. The population is currently estimated to be around twenty thousand, and around half are migrants.[27] Since the beginning, the indigenous population has been against the mining exploration, as these locals perceive the area as sacred place. The first conflict occurred in 1967 when the initial exploration started. Resentment against Freeport began for the second time in 1973 and in 1977 when a larger conflict occurred, which involved the rebellious group of OPM. Ever since this second conflict, the military began to be deployed in Timika, which further inflamed resistance from the local population. The perception that Freeport only benefits the foreign investors and the central government has been the major source of resentment. Furthermore, the atrocities committed by the military on the local people and various human rights violations increased the potential for conflict in Timika. The local resentment toward Freeport and the military culminated in several riots in 1995 that become national and international issues.[28]

Population mobility and employment data as revealed from population censuses and labor force surveys during the last two decades clearly indicate that the mode of population mobility was strongly associated with the broader social–economic processes instigated by government policies. The increase in the volume of population mobility, as mentioned earlier, is essentially the logical consequence of three factors: first, the surplus of labor; second, the improvement of transportation means and networks; and third, the opening up of economic activities, particularly in urban areas where the informal sector economy provides many different job choices for migrants.[29] In Eastern Indonesian provinces, the most prominent migrants, besides the Javanese—who are almost everywhere, are the three ethnic groups from the southern parts of Sulawesi, namely the Bugis, the Makassarese, and the Butonese. These migrants are well known as seafarers and traders and have successfully

[26] On migration to East Timor, see Soewartoyo (1994), and Aditjondro (1994).

[27] See Widjojo (1995).

[28] For a detailed account of the conflict between the indigenous people (Amungme and Kamoro) and Freeport Inc., see part four of a book on human rights violations in the mining industries in Indonesia (Bachriadi 1998).

[29] Since the 1980s, large numbers of migrant workers have also moved across the national boundaries, particularly to Malaysia.

developed their social and economic networks in major urban areas in Eastern Indonesia.

Politics of Migration: East Timor and Irian Jaya

In Eastern Indonesia, specifically in Irian Jaya and East Timor, the division of labor by ethnic groups is somewhat different in comparison with other places.[30] The upper-level economic sectors are usually in the hands of the Chinese, whereas the middle and the lower levels are dominated by the Makassarese, the Bugis, and the Butonese. The Javanese, including other migrant–ethnic groups, usually occupy limited government jobs, such as public servants and the military. The Javanese domination of various governmental occupations is a by-product of the process of employee recruitment, which is conducted by the central government in Java. In Irian Jaya and East Timor, due to their political histories, the appointment of officials by the central government is also dictated by security and military considerations. The feelings of resentment among the local population toward the domination of migrants in both governmental and private occupations are obvious and easily perceived, even by a casual observer.[31]

Given their distinct sociopolitical circumstances, conflict between migrants and local populations in Irian Jaya and East Timor cannot be understood if related only to demographic and economic issues. In assessing the events, the political histories of the provinces—in their relation to the nation-state building processes—should be given special consideration. Irian Jaya Province, formerly West Papua—a Dutch colony, was integrated into the Indonesian state in 1963 after long diplomatic negotiations and incidences of armed conflict, between Indonesia and the Dutch. Irian Jaya was incorporated into Indonesia after the state won a referendum under UN supervision. However, a group of West Papuans rejected the outcome of the referendum and proclaimed their organization (*OPM*) as the legitimate owner of the province and took up armed struggle against Indonesia. East Timor was annexed by Indonesia in 1976 amid internal conflict between political parties competing for their legitimacy to rule the former Portuguese colony.

Similar to Irian Jaya, armed struggle against Indonesia's occupation existed, under the banner of *Fretilin (Frente Revolucionária de Timor-Leste Independente)*.[32] Despite Indonesia's strong claim to the territory, the UN still

[30] Information on the division of labor by ethnicity was observed on the author's intermittent visits to Eastern Indonesia between 1991 *and* 1993.

[31] The migrants are pejoratively called by the locals, the "straight-haired," while the migrants call the locals the "curly haired."

[32] On the impact of the Indonesian occupation of East Timor, see Soetrisno (1991) and Aditjondro (1994).

recognizes the Portuguese as the administering power of East Timor.[33] Despite the volatile political climate, the central government has continuously provided the province with substantial budget allocations, most notably to develop various physical infrastructures. The construction of asphalt roads linking West and East Timor, and the creation of regular flights and ships to and from Dili—the East Timor Capital, significantly contributed to the increase in in-migration to East Timor. An underpinning demographic factor in the building up of tension between migrants and local people has been the burgeoning of unemployment, particularly among the educated youth in urban areas. In 1990, for example, in Dili, the capital of East Timor, more than 20% of the labor force between the ages of 15 and 29 were unemployed.[34] In most cases, the involvement of local youth in the various forms of protest the migrants was very prominent. Furthermore, the role of the underground resistance movement within East Timor against Indonesia's occupation should not be underestimated in assessing the conflicts between in-migrants and the local population. In East Timor, migrants often decided to leave the province to avoid the harsh treatment by the East Timorese. The exodus of migrants, many of them are *the BBM*, from East Timor occurred in September 1996 and July 1998, following a large anti-integration protest.

The power equation between the center and the regions after the downfall of Suharto on May 21, 1998, was shattered as the various groups protesting the central government domination began to flourish. The demand for more autonomy has been most clearly articulated by the East Timorese and Irianese people. The call for a referendum that was strongly suppressed during the Suharto regime is now being loudly expressed, most notably by the anti-integration student and youth organizations. The new government, under President B.J. Habibie, has responded positively to the demand for more autonomy in East Timor. At the last tripartite meeting, with his Portuguese counterpart Jaime Gama under the auspices of the UN Secretary General, Kofi Annan, on August 3, 1998, in New York, Minister of Foreign Affairs Ali Alatas, promised to provide "a wide-ranging autonomy" for East Timor.[35] At the same time, the demand for autonomy is also increasing in Irian Jaya, which has led, for example, to the formation of the Reconciliation Forum of the People of Irian Jaya (*Foreri*), which comprises three groups, respectively, calling for "an autonomous state," "an independence state," and "a federal state."[36]

Another development occurred in the westernmost provinces, Aceh, which has long experienced political repression from the central government, due to the accusation that the *Gerakan Aceh Merdeka* (Free Acehnese Movement) is still

---

[33] As a result of its disputed status in the international forum, East Timor, in any case, therefore, warrants special consideration. The rise of the underground movement, interestingly organized by young people, obtained more international attention, after the so-called Dili Incident on November 12, 1991, in which an estimated 100 people were murdered by Indonesian soldiers while conducting a protest rally. Since 1991, East Timor obviously has become the most troubled region in Indonesia.

[34] See Tirtosudarmo and Handayani (1993).

[35] Jakarta Post, August 3, 1998.

[36] Jakarta Post, August 3, 1998.

operating in this province.[37] The Aceh situation is perhaps rather different from East Timor and Irian Jaya: The issue is more concentrated on human rights violations, particularly by the military than a secessionist movement.[38] In an unprecedented move on August 7, 1998, the Minister of Defense and Security/Armed Forces Commander (*ABRI*) General Wiranto pledged to withdraw troops deployed in Aceh and apologized for misconduct by soldiers while stationed there. Aceh, Irian Jaya, and East Timor are the provinces that were unofficially declared as the military operation areas (*Daerah Operasi Militer* or *DOM*), mainly because in these three provinces, the military perceived them as areas in which armed separatist groups fighting for an independent state still exist.

Concluding Remarks

Migration in Eastern Indonesia, during the New Order regime, reflects the regional response to the wider economic and political changes, particularly the centralistic character of the New Order economic development. Similar to the national migration pattern, in-migration has been declining since the mid-1980s. Since that time, migration toward urban areas has been increasing. This tendency is partly because the number of transmigrants, who are mostly resettled in rural areas, is declining as the government capacity to finance the transmigration program has diminished. The migrants to urban areas in Eastern Indonesian provinces are mostly attracted by the small trade and informal sectors—which are becoming competitive economic public spheres among the migrants and local or indigenous people. While the brightest local people have migrated to Java or other areas in Western Indonesia, most locals who are left are therefore less educated and less skilled in comparison with the migrants.

The centralistic character of the national government, both civilian and military, in which regional-level bureaucracies are merely subordinates of the central government, resulted in a government public servant recruiting process that is heavily dictated by the central government. This centralistic recruitment process obviously resulted in the appointment of bureaucrats who are mostly outsiders from the perspective of the local populations.[39] The unofficially declared military operations areas, which led to the excessive presence of military personnel, in Irian Jaya and East Timor, further exacerbated the feelings of being colonized by the New Order regime among the local populations. The feeling of being economically

[37] For a detailed account of the impact of the military operation in Aceh, see ELSAM (1995, 13–31).

[38] Reported atrocities by the military in Aceh, Irian Jaya, and East Timor, mostly by NGOs and religious groups, including abductions, rapes, torture, and extra-judicial killing, are flourishing in the mass media since the resignation of President Suharto on May 21, 1998, and the prevailing new climate of openness.

[39] The centralization of the public servant recruitment process also reflected the Javanese cultural mobilization of the Outer Islands and the periphery in general—an important part of the nation-state building strategy (Magenda 1986).

marginalized and politically suppressed is latent in Eastern Indonesia, particularly in Irian Jaya and East Timor, and less severe in Maluku. In the New Order period, given the political histories of Irian Jaya and East Timor: Therefore, migration has becoming politicized, and it would be misleading if it were perceived merely as being demographic or economic.

In the current heated political climate, the issue of regional development can no longer be isolated from the debate surrounding regional political autonomy and the demand to reformulate the existing national state formation. The emergence of various political parties demanding a free and democratic general election, which is promised by the B.J. Habibie government to be conducted in June 1999, is likely to be interconnected with the demand for more autonomy and independence from the regions, most explicitly East Timor and Irian Jaya. The result of the current ongoing process of the formulation of new laws on political parties and general elections, for example, will therefore be crucial and depend on whether the new national political format can accommodate the strong call for regional economic and political autonomy. The planning for regional development in the future, therefore, should take a more comprehensive approach in which the interlinkages between the achievements of economic welfare should be accompanied by the fulfillment of the political aspirations and human rights of the local people.

Chapter 6
West Kalimantan: The Political Demography of Borderland

Border in the Center's Perception

Since the beginning of the New Order era, the obsession concerning national integration and modernization has obtained its perfect form through various policies and programs designed by its two main proponents, namely the military elite and economists–technocrats. Both the military and the economists–technocrats, as the main actors representing state power, described community in a similar way. The community, in the militaries and economists–technocrats' perspective, was as follows: first, a group of human beings having uniformity in their various characteristics; and second, such group of human beings occupies a geographical space viewed as a flat surface. Based on this perception, the military and the economists–technocrats treated the community as something that could rationally and systematically be designed and engineered from a certain control center in the desired direction. Differences and cultural diversities inherent in the community were deemed nonexistent or irrelevant in social engineering and development planning. Various policies and programs were uniformly drawn up for the entire territory of Indonesia and were implemented through centralized control. Development planning was based only on the statistical figures of economic demography and various aggregate economic indicators.

In the meantime, the military elite who considered themselves as the party that had inherited the fighting spirit of freedom fighters perceived themselves as the only group having interests in maintaining territorial unity and national integration. During its development, especially during the New Order era, the military created an organizational structure providing a great opportunity to control and mobilize the people for the sake of maintaining the integrity of state territory. The People's Security and Defense System *(Sishankamrata)* serves as the principal guideline in the working mechanism of the military organization, materialized in the form of the territorial command system, the jurisdiction of which covers the entire territory of Indonesia (Sebastian 1996). While the economists–technocrats worked based on the

R. Tirtosudarmo, *From Colonization to Nation-State*,
https://doi.org/10.1007/978-981-16-6437-3_6

perceived growth and development, on the other hand, the military acted based on perceived threat. These two perceptions, in addition to being mutually supplementary in nature, also contain a concept of a "socially and culturally homogeneous community."

Cultural pluralism and heterogeneity in the Indonesian community were indeed beyond the imagination of the military and economists–technocrats who were immensely powerful during the New Order era. The ethnical discourse reconstruction process during the New Order era culminated in the SARA (Ethnic Affiliations, Religion, Race, and Societal Groups) concept. The SARA concept, which subsequently became the main reference for state policies, basically conceived a harmonious community without any conflict. The concept implied that differences and diversities were deemed as sources of conflict that had to be avoided. The ethnic heterogeneity inherent in the Indonesian community, therefore, had to be merged through various policies and programs producing the national culture and characteristics, which constitute the identity of the Indonesian people. The ethnic heterogeneity at this level of comprehension was the source of ethnic conflicts.

The dynamics of political demography occurring in West Kalimantan cannot be separated from political and economic changes occurring at the national level, both in the past and at present; the changes are reflected in the center's perception of the border area. In this chapter, the social and cultural dynamics of border area will be viewed from the perspective of political demography.[1] Political demography is an approach considering a social or cultural problem as the consequence or implication of the interaction or reciprocal relations between demographic changes and political changes. Observing the social and cultural dynamics in West Kalimantan and Sarawak border area from the perspective of political demography means trying to explain various social and cultural issues arising in the border area. This is the consequence and implication of interactions of demographic and political factors in West Kalimantan and Sarawak border area and its vicinity.

To explain the social and cultural dynamics in the border area viewed from the perspective of political demography, this chapter is generally divided into four parts. Part I constitutes the description of West Kalimantan from the viewpoint of social demographic changes, especially in its relation to the growth, spread, and composition of its population. Part II contains my observation of two peripheral locations

[1] The demographic dimension or aspect, in an analysis of various social problems, is often deemed merely as background information that does not have any significance on the theoretical comprehension of social sciences and alternative strategies offered to solve social problems confronted. This type of perspective is developed because of the mistakes of the demographic experts who in their historical process have made demography an exclusive discipline that tends to close itself to interaction and cross-fertilization with other disciplines of social sciences. This introversion tendency is also caused by the closeness of demography to various practical public policies, especially in its relation to various global population growth control programs. At present, demographic experts and persons interested in demography gave beginning to be aware that demography as a scientific discipline will experience an impasse and stagnation if they remain only as a part of public policy. This policy is a social engineering tool that often has a destructive nature to the future of humanity.

between West Kalimantan and Sarawak, namely Jagoi Babang and Entikong. In Part III, I discuss the political changes affecting sociodemographic dynamics in West Kalimantan, especially in their relation to the geostrategic position of West Kalimantan as an area bordering on another country. Part IV contains my explanation on the two main perceptions of the center (perceived threat and perceived advantage) of the "border area."

Migration and Social–Demographic Changes

West Kalimantan is one of the provinces in Kalimantan Island,[2] which directly bordering on Sarawak, one of the states in East Malaysia. In such a position, West Kalimantan can have direct contact with a foreign country. Such direct relationship is made possible by the opportunity for Indonesian people to directly cross the border to Sarawak without having to cross the sea and through ongoing trading activities between Indonesia and Malaysia. These trading activities are carried out through land transportation facilities crossing the West Kalimantan border area. Human traffic and trading activities to and from West Kalimantan and Sarawak—that have existed for quite a long time—have also contributed to shaping and influencing social–cultural dynamics in the border area.

According to the results of population survey between censuses (SUPAS) in 1995, the population of West Kalimantan was 3.6 million. The growth of West Kalimantan population was quite high, averaging approximately 2.4% per annum from 1980 to 1995. This high growth was mainly due to the large number of migrants entering West Kalimantan in the said period. Although West Kalimantan is deemed as having a high population growth considering the size of its area (146,670 km^2), there is only an average of 25 people per kilometer. Compare this to Java Island having more than 850 people per kilometer or Jakarta which have almost 14,000 people per kilometer. Seen from such a demographic perspective, we can easily conclude that West Kalimantan is a sparsely populated province or even an empty area.

The perception and assumption that West Kalimantan or Kalimantan Island is an "empty" area substantially affect the stances and actions of decision makers in Jakarta in respect of West Kalimantan and Kalimantan Island in general. One of the persistent stances is the desire to relocate the people from densely populated areas such as Java, Madura, and Bali to West Kalimantan and Kalimantan Island in general. This stance is subsequently realized in the form of various government

[2] Kalimantan, as a name of an island, is used in Indonesian language. Borneo is another name of Kalimantan used outside Indonesia. In his explanation, King (1993, 18) said that "the Indonesians use it specifically to refer to the southern two-thirds of the island, but it is also widely accepted in Indonesia as a general name for the whole island. The association of the term 'Borneo' both with the non-Indonesian Sultanate of Brunei and with Western corruption of a local name must have played a part in its rejection by the strongly nationalistic Indonesians."

actions and policies, such as the colonization program in the colonial era and the transmigration program in the postindependence era. In addition to the stance of the ruling elite—both prior to and after the independence—considering West Kalimantan as an empty area, the economic potentials and natural resources contained in West Kalimantan have been especially attractive for spontaneous migrants. They have come from various areas outside Kalimantan and have migrated to this area in pursuit of their fortune.

The history of West Kalimantan can be described as a history of migration or in other words, a history of interaction between migrants and the local people (or those who had arrived earlier) in this area.[3] As it is situated on the intersection of Malaya Peninsula and Java Island, also of Sumatra and the Philippine archipelago, as well as Sulawesi, West Kalimantan, and Kalimantan Island have become a transit place of various ethnic groups throughout Indonesia in general. It is interesting to note that, evidently, Borneo or Kalimantan has a unique meaning for the Europeans as expressed by King (1993, 28) "For a long time much of Borneo remained at a distance from European interest and adminis-tration—a neglected and relatively unknown is-land." It entered European thought and writing as a hostile place: inhospitable, dangerous, bar-barous, and difficult to govern.

In West Kalimantan and Kalimantan Island in general, the settlements of the "indigenous people" are spread in small groups, and each of them has different characteristics. The Dayak people, however, are generally categorized as "the native's people of Borneo."[4] Dayak people as a general category include groups of people such as Iban, Kayan, Bidayuh, Kendayan, Maloh or Taman, Lun Bawang, and the like (King 1993, 29). The nomenclature of the groups has its own history and has always been problematic. In Kalimantan, Dayak people and Punan people have always been compared to the Malay people.[5] The contrast between Dayak and Malay people is based on their religion. Malays are always associated with those who profess Islam, though actually the Malays can also be differentiated based on their language or customs and traditions. The Malays are very heterogeneous, though most of them are natives who subsequently professed Islam. Concerning this, King (1993, 31) explains as follows "Gradually over time religious conversion would result in a reclassification of the Dayak converts as Malays; the local Malay term for this pro-cess is masok Melayu ('to become Malay' or 'to enter Malaydom') or sometimes turun Melayu ('to come down [and become] Malay')."

---

[3]Tensions and conflicts between migrants and local people are an important part of the social and political history of West Kalimantan. Claim as local people or "sons of the soil" as used in political discourse is a classic issue that always rises in any political unrest in West Kalimantan.

[4]In the literature, the Dayak people are also called "non-Malay natives," or some differentiate land Dayaks (Bidayuh people) mostly occupying the western part of Sarawak from sea Dayaks (Iban people) spread in coastal areas. Iban people and the Malays, during the governance of James Brooke in Sarawak, were notoriously known as scary pirates (King 1993, 29–30).

[5]Ethnolinguistically, Collins (1999) indicated the Dayak people and the Malays in West Kalimantan from one family.

According to King (1993, 31), there are many Dayak groups, such as Kendayan, Bidayuh, and Maloh people, who have become Moslems and subsequently called themselves Kapuas Malays. The Malay term is also frequently used as a suffix, such as Banjar Malay, Brunei Malay, or Sarawak Malay. In the subsequent development, Malay groups became larger with the entry of Moslems from Java, Sulawesi, or the Malay Peninsula. The Malays generally live in coastal areas or along the rivers, which subsequently became markets. The majority of the Malays are engaged in small-scale trading activities in addition to fishery and rice field cultivation, as stated by Warren (1981): namely that the main economic activity in Kalimantan is not agriculture but trading and taxes. King[6] (1993, 27) says:

> Trade and tributary relations had existed be-tween coastal states and hinterland population from at least the middle of the first millennium AD, and it was not the case that head-hunting always operated in a separate political sphere from piracy. Some of the headhunting tribal groups, and especially the Ibans, were recruited as mercenaries by Malay coastal raiders.

Javanese began to enter Kalimantan in the fourteenth and fifteenth centuries during the glorious period of the Majapahit Kingdom, which then controlled a part of Kalimantan Island. The number of Javanese rapidly increased since the New Order Era owing to the transmigration program. Many Madurese are also found in Kalimantan, and they live in urban areas. They work as daily laborers or pedicab drivers. They also engage in plantation, fishery, or animal husbandry. The number of Moslems is estimated to reach around 60–65% of the entire population of Kalimantan (King 1993, 33). According to a source, in 1991, 54% of the West Kalimantan population was Moslems, whereas 9% were Protestant, 19% were Catholic, and the rest were Hindu, Buddhist, or followers of other religions (Frederick and Worden 1993, 353).

In addition to Dayaks, Malays, and Moslems from other islands, there are also Hindu migrants from Bali, and Christian migrants from Eastern Indonesia, such as from Timor and Flores. They generally work in the agricultural sector. Overseas migrants mostly came from China, and a few came from India and Arabia (who then mingled with the Malays). Currently, the Chinese live in urban areas and are engaged in trade and finance like the Indians. The Chinese had first come to Kalimantan in the eighteenth century in a quest for gold. As the gold mining ended, the Chinese in the western part of Sarawak and West Kalimantan turned to the agricultural sector.

Based on the population census and population survey between censuses (SUPAS), Agustiar (1998) notes that as from 1971 up to and including 1995, the volume and flow of incoming and outgoing migration in West Kalimantan indicate interesting changes. First, the increased number and percentage of lifetime migration from areas other than Java Island. This condition has proven that there were more areas of origin of migrants to West Kalimantan, though migrants from Java were still

[6]It was trade, tribute, tax, and toll collection and not agriculture that provided the raison *d'être* for the expansion of state formations in Borneo (Warren 1981, xxii–xxvi in King 1993, 24).

dominant.[7] Second, the largest outgoing migration from West Kalimantan for 25 years was to Jakarta. It is assumed that the migration flow to Jakarta from West Kalimantan was dominated by people from Pontianak Chinese many of whom were popularly known to engage in various trading businesses in Jakarta. In addition to that, there were also young people from West Kalimantan intending to continue their education and looking for jobs.

The incoming migration to West Kalimantan other than from Java was originated from North Sumatra, West Sumatra, Riau, West Nusa Tenggara, and South Sulawesi. According to Agustiar (1998), migrants from North Sumatra entered West Kalimantan because they are attracted by large plantation estates (PTP) associated with the plantation estates in North Sumatra. Migrants from West Sumatra and Riau came as the result of "chain migration" by previous migrants from the two provinces, while migrants from East Nusa Tenggara arrived as a result of the mass assignment of teachers in the 1980s to various remote areas in West Kalimantan. Furthermore, migrant workers were also attracted by job opportunities in timber industries that experienced a booming following the 1970s, not only in West Kalimantan but also in Kalimantan Island in general including Sabah and Sarawak. Migrants from South Sulawesi were sailors dominating the coastal areas. Many of them then opened coconut plantations.

Agustiar (1998, 147) explains the four general patterns of migration in West Kalimantan. First, migration occurring in the past (the seventh century) marked with the arrival of many commercial associations from China, especially people of the Hoakiau ethnic group from Fukien and Hakka ethnic group from Kwantung. Both ethnic groups came to West Kalimantan to work in gold mines in Montrado, Sambas, and Bengkayang regions. Second, migration by Bugis, Banjar, and Madurese, which among other things, is evident from the name of villages, canals, and rivers related to the names of ethnic groups of the said migrants such as Banjar Sarasan Village, Dalam Bugis Village, Bugis Channel, and Wa''Daeng Channel. Third, after the independence, the migration pattern was indicated by the migration of retired soldiers (people's army) and their families from Java Island. Fourth, organized migration from Java Island in the context of transmigration programs. Fifth, migration to the neighboring country (Sarawak and Sabah)—either legally (through gate point Entikong/Tebedu), or illegally through the numerous of "dirt trails" bridging West Kalimantan and Sarawak. This type of migration has been the most modern pattern of migration and the most relevant to the social–cultural dynamics at the border area.

[7] Compared to the year 1971, the percentage of the incoming migration in 1995 decreased from 77 to 75% (Agustiar 1998, 133).

Observation of Two Borderland's Sites: Jagoi Babang and Entikong

The intentionally selected observation conducted in Jagoi Babang, due to its "dirt trail" connecting West Kalimantan and Sarawak, clearly indicates a free flow of people and goods between the two countries.[8] The flow of people and goods took place without immigration or customs procedures as usually required in interstate traffic.[9] It is interesting to note that there has been an asphalt road that can accommodate four-wheeled vehicles (trucks or minibusses) connecting Pontianak and Jagoi Babang. In addition to that, a customs office and immigration office had also been established, which seemed not to be functioning.

According to a source, there has been a plan to make Jagoi Babang an official entry and exit gate between West Kalimantan and Sarawak. The Indonesian Government has constructed infrastructures for such purpose, among other things, asphalt road connecting Pontianak with Jagoi Babang and building facilities for immigration and customs authorities. On the other hand, the Malaysian government has not done anything yet to reciprocate Indonesian efforts. On the Malaysian side, for example, there were no roads like those provided by Indonesia.[10] Trucks carrying goods from Pontianak had to stop in Jagoi Babang and the transport to Malaysia then continued the trips by motorcycles and bicycles because the roads to be taken were footpaths through forests. A source in Jagoi Babang said that a trip to a "village" and "market" through a foot path would only take 30 min by motorcycle. The illegal cross-border goods transactions were then conducted in the village concerned. In such transactions, cigarettes, *Indomie* (instant noodles), as well as various snacks (biscuits), and canned drinks (beer, soft drink, and so on) were traded from Sarawak. In addition to the traffic of goods, there were also young women carrying large bags or sacks entering Malaysia. They were female workers or job seekers whose number reached thousands in Sarawak. The increased number of job seekers in Sarawak was estimated to have occurred due to the impacts of the economic crisis hitting Indonesia since the middle of 1997.[11]

---

[8] Visits to the border areas were made twice during the research. The first visit was made in August 1999 to Jagoi Babang, Sambas Regency. The second one was made in August 2000 to Entikong, also in Sambas Regency.

[9] According to some sources in West Kalimantan and Sarawak, the dirt trail found in Jagoi Babang is one of the tens or possibly hundreds of "dirt trails" existing in the borderlines between West Kalimantan and Sarawak.

[10] A source said that Malaysia became more cautious in developing transportation facilities with Indonesia with the occurrence of ethnic unrest in West Kalimantan in 1996, 1997, and 1999.

[11] As expressed by the Indonesian Consul in Kuching, the number of violent cases especially toward housemaids coming from Indonesia is increased on the brink of a visit to Kuching by the research team in February 2000. Unintentionally, when the research team visited the Indonesian Consulate in Kuching, around 30 male plantation laborers from East Flores were being accommodated at the backside of the consulate. They admitted that they were not paid by worker providers in accordance with the work contract they signed. When the research team paid a visit, they were waiting for the

Unlike Jagoi Babang, which represents illegal gateways between West Kalimantan and Sarawak, Entikong is a legal gate connecting Pontianak and Kuching—the capital of Sarawak. Dozens of buses carrying passengers passed to and from Pontianak and Kuching every day, taking approximately 9 h. The frequency of the traffic of private vehicles and trucks carrying goods to and from the two countries was also high. This made the immigration and customs authorities in Entikong and Tebedu (on the side of Sarawak) quite busy. There was an interesting phenomenon namely the emergence of an illegal market in Indonesia, which was not far from the Entikong entry point, a place where various types of goods (garments, electronics, foods, rice, and so on) were freely traded. This illegal market appeared as the result of the depreciation of rupiah value against the Malaysian ringgit. Traders established illegal kiosks next to the entry point facilitating buyers from Malaysia to spend their ringgit on low-priced goods. In addition to the illegal market, the drop of the rupiah values also spurred street money changers nearby the Entikong entry point. The economic crisis hitting Indonesia opened various new job opportunities in border areas.

Political Changes and the Implications Thereof

Kalimantan is a province, which has a very heterogeneous ethnicity composition compared to other provinces in Indonesia. The ethnic heterogeneity in West Kalimantan mainly resulted from the inflowing migration of various ethnic groups from regions other than West Kalimantan. Besides being heterogeneous, West Kalimantan is also a province with the largest number of Chinese ethnic groups. A source said that the population of the Chinese ethnic group in West Kalimantan reaches around 11–15%; meanwhile, in Indonesia the number is around 2–3%.[12] As explained in the first part, the natives of West Kalimantan are Dayaks, who are divided into many subethnic groups which means that they do not constitute a homogeneous group. Based on the 1930 population census, the number of Dayaks in West Kalimantan was 42%; meanwhile, Malays constituted 29% (Mackie 1974, 349 in Davidson 2000, 4). A rather drastic change occurred in 1971: The percentage of Malays increased to 39%, while Dayaks decreased to 41% (Sukarnaprawira 1975, 90 in Davidson 2000, 4). In the meantime, Javanese, Bugis, and Madura people were estimated to be around 3%, respectively.[13] In Pontianak, the Chinese ethnic group

negotiation result between the employers and the worker providers. They demanded that their salaries be paid following the contract and returned to East Flores.

[12]This is a raw estimation because there is no source of data on ethnicity in Indonesia; consequently, different sources have different estimation figures.

[13]The number of Madurese in Pontianak possibly underwent a significant increase following the occurrence of ethnic unrest involving the Madurese: first with Dayak people in Sanggau Ledo (December 1996 and March 1997) and second with the Malays Sambas Regency (February–April 1999). As the result of such ethnic conflict, the Madurese had to flee to various places, among other

was estimated to constitute around one third of the total population, whereas Malays constituted one quarter, Bugis 3%, Javanese 12%, and Dayaks 3% (Davidson 2000). As it borders on a neighboring country, West Kalimantan has a "strategic role" viewed from the aspect of state defense and security interests. For the people in West Kalimantan, especially those living in border areas, the confrontation with Malaysia occurring during the era of Sukarno's administration has greatly affected their perception of themselves, the state, and the neighboring country, namely Sarawak–Malaysia. According to Mackie (1974, 347–348 in Davidson 2000, 5), West Kalimantan had started to attract the attention of the central government at the beginning of the 1960s when Malaysia proclaimed its independence. Sukarno considered Malaysia as an English puppet state constituting a threat of Western imperialism. He opposed the establishment of Malaysia and launched confrontation politics. In 1963, West Kalimantan served as Indonesian military base in the confrontation with Malaysia. At that time, West Kalimantan had entered the Indonesian political map.

The confrontation with Malaysia ended in August 1966, but its implications continue to date, especially in the vicinity of border areas, in the form of bloody conflicts between Dayaks and the Chinese inhabitants in October 1967, "to inscribe its authority in the region, the In-donesian military, under the command of the New Order's territorial military leadership—Kodam Tanjung Pura/XII—investigated and facilitated a massive slaughter of ethnic Chinese in West Kalimantan. Military authorities insisted that the centuries—Old Chinese village near the Sarawak border were being used as cover and supply bases by a predominantly leftist Malaysian Chinese paramilitary organization—the Sarawak People's Guerrilla Forces (Pasukan Gerilya Rakyat Sarawak or PGRS), who were ironically trained by the Indonesian army to help 'de-stabilize' Sarawak during confrontation" (Davidson 2000, 5).

Since then, by mobilizing Dayaks, Chinese people were forced to abandon their villages and moved to towns by the military. The reason quoted by the military was to facilitate the Military's control over Chinese inhabitants. This period lasted for 3 months, known in West Kalimantan as the demonstration period. At that time, thousands of Chinese inhabitants were killed and thousands more fled to coastal areas in the western part of West Kalimantan, especially to Pontianak and Singkawang. Furthermore, with the region being "secured," New Order authorities and cronies, armed with authority from Jakarta and international capital and backed by new capital-friendly foreign investment and forestry laws, immediately began to log the province's immense forests. The Basic Forestry Law, Law No. 5/1967, which claims the country's forests as state property, set up the framework for the granting of the large-scale concessions under the Forestry Concessionary System *(Hak Pengusahaan Hutan,* or HPH).

Since then, West Kalimantan has become a cash cow for various interest groups in Jakarta, both civilian and military. For example, in 1993, the central government issued a regulation to control orange trading in West Kalimantan, which was

things, to Pontianak. The government accommodated them for the time being in a sport stadium. The number of Madurese living in the settlements is estimated to reach more than 25,000 lives.

extremely harmful to local orange farmers because it was a form of orange trade monopoly in West Kalimantan. As a result of that, in 1996 many orange farmers went bankrupt and some of them subsequently decided to move to Sarawak to work on plantations. Exploitation by the central government in West Kalimantan resulted in the marginalization of the local community in social, economic, and cultural aspects.[14] The gist of this massive marginalization process was the elimination of rights on land, which were previously recognized by the local customary law. However, today such rights have been eliminated by the issuance of new laws during the New Order era, such as the Agrarian Law, Forestry Law, and Mining Law. Those laws legitimize the state's control over land previously owned by customary groups (Djuweng 1997, 22, in Davidson 2000, 8).

At the beginning of 1980, the central government began to relocate people from scattered hinterland to permanent and integrated settlements. According to the government's evaluation, 77% of the 4362 villages in West Kalimantan were occupied by less than 100 households. Through the village rearrangement program, sparsely populated villages were unified resulting in 1297 new larger villages (West Kalimantan in Figures 1997, 41 in Davidson 2000, 8). The motivation of the government's program in merging hinterland villages, which mostly were occupied by Dayaks, was to support the "development," and "modernization" program that was the credo of the New Order regime. In addition to "development" purpose, this village rearrangement program was also spurred by "strategic military" considerations to make control and mobilization of people for the political interests of the New Order easier, among other things, to help *Golkar* to win the General Elections.[15]

The village arrangement program in West Kalimantan has significantly changed its local leadership pattern and structure. Customary leaders, who used to be powerful, have been drastically deprived of their power by the state. Their power has been transferred to village chiefs or hamlet chiefs administratively constituting the state accessories. Since then, customary leaders have only had symbolic powers, especially to handle moral problems such as violation of the custom and tradition related to marriage or social conflicts. These problems are also handled by state agencies, such as the police. Customary law is practically difficult to be applied in a community that has experienced an enormous social transformation like those in West Kalimantan.[16]

---

[14]Concerning population marginalization process during the New Order era, see Chap. 10 "Modernization and Development" in King (1993), and Alqadrie (1994).

[15]It is not surprising when *Golkar* won in almost all rural areas in West Kalimantan in the 1999 General Election, whereas *PDIP* deemed as a reformist party won only in Pontianak Municipality (Davidson 2000).

[16]In the fieldwork, the writer interviewed some customary leaders in Sonta Village, not far from Entikong. Sonta Village occupied by Bidayuh Dayak ethnic group constitutes a village being the unification of various spread-out population settlements. The Bidayuh people live not only in West Kalimantan territory but also in Sarawak territory. Some of those who live in West Kalimantan and Sarawak still have family relations.

West Kalimantan has been the main destination for transmigration since the early 1970s. Prior to the launching of the transmigration program by the New Order administration at the beginning of the 1960s, West Kalimantan had been used as the destination for the resettlement of retired servicemen. Those former servicemen were placed nearby the border with Sarawak with the purpose to improve defense and security as part of the confrontation politics of Sukarno's regime at that time. Statistically, the number of transmigrant settlements in West Kalimantan improved. In 1971, the number of transmigrants was only around 0.5% of the total population, which subsequently increased 14-fold in 1985, (Fasbender and Erbe 1990, 137 in Davidson 2000, 9). This phenomenon of increased population number is described by Davidson (2000, 9) as follows:

> The state has filled this politically constructed empty space with more "manageable" and pliable transmigrants from Java and Madura. Although a part of West Kalimantan popu-lation history was a natural or spontaneous de-mographic process, it is difficult to deny that most of it constitutes a demographic engineer-ing process. The engineering was conducted by the state since the colonial era up to the New Order era as put forward by Davidson (2000): the state's success at reconfiguring the province's population, however, has pro-duced greater problems than anticipated.

Open interethnic tensions and conflicts often occurred in West Kalimantan,[17] among other things, conflict between Dayaks and Madurese reaching its peak in December 1996 to March 1997 in Sanggau Ledo District, Sambas Regency.[18] This conflict resulted in thousands of Madurese had to abandon their residences. In February–April 1999, an ethnic conflict reoccurred in Sambas Regency. The conflict was between Madurese and Malays, which happened for the first time. Because of that event, around 25,000 Madurese had to seek refuge in Pontianak. Unlike refugees of the Dayaks–Madura conflict in 1996/1997 who could return to their residences in Sambas, this time the Madurese could not return, and they had to live in various refugee camps, especially in Pontianak.[19]

On October 27, 2000, the mass media reported that (it was even televised on CNN) a riot had occurred, which was initiated by a car accident on Wednesday, October 25, 2000, at 8 p.m. on a road in Pontianak involving a Malay and a Madurese.[20] Pontianak was full of tension at that time because the disputing groups gathered in quite large numbers and in a ready-to-fight condition. The security apparatus managed to obstruct both groups from meeting each other. Later, the governor eventually imposes a curfew. Like the previous interethnic conflicts, this

[17] A Chinese uprising occurred in West Kalimantan in 1914 (see The Siaw Giap 1981).

[18] According to a source, conflicts between Dayak and Madura had once occurred in 1968, 1976, 1977, 1979, 1982, 1983, and 1992 (Human Right Watch 1997).

[19] Concerning Madura–Melayu conflicts, see Davidson (2000).

[20] Generally, the printed mass media reporting this incident did not mention the ethnic groups involved. The inclusion of Madura and Malay ethnic groups was obtained from *Astaga*.com, Wednesday, October 25, 2000, "Pontianak Unrest." In ethnical upheavals claiming many casualties, the security apparatus (police and army) almost always failed in securing the condition so that they developed to become a killing field, usually on the party of the lesser number.

upheaval was also initiated by an individual's agitation or dispute between two persons from different ethnic groups, which subsequently quickly developed into a riot involving large masses of the different ethnic groups.[21]

The relationship between population composition based on ethnicity and the occurrence of open interethnic conflicts, such as those occurring in West Kalimantan, does not occur in a vacuum situation. Tensions among different ethnic groups are a process taking place gradually but accumulatively. The perception of the differences between one ethnic group and the others solidifies. Number and balanced composition of population based on ethnic identity are crucial problems. Based on the observation and publicized statistical data, the larger the number of Malays, Madurese, Javanese, and Bugis people, and like in West Kalimantan, the more it causes a sense of anxiety and oppression. The Dayaks are also encouraged to resist and fight. The composition of population based on ethnicity often triggers "the politics of numbers," as occurring in Malaysia between Malay ethnic and Chinese ethnic groups whose numbers are almost balanced. At the global level, the increased number of immigrants from the poor—south countries to rich—north countries has become a heated political issue. This is based on the perception of the fear of change in the population composition and number balance,[22] as described by Esman (1994, 13–14):

> Ethnicity as politics is meaningful only in a re-lational framework. As a collective identity is shaped not only self-definition but also with con-straints imposed by outsiders. Thus, the internal "we" must be distinguished from the external "they." Where no relevant others exist, the need for solidarity disappears and society fragments based on internal differences. The need for ethnic solidarity arises only when strange, threatening, competitive outsiders must be confronted. Only then are distinctions between "us" ad "them" noted, explained, and evaluated to reinforce internal coherence and to signal and explain differences with outsiders and competitors. Only then do the needs for internal solidarity and discipline become evident.

Human Right Watch (1977), in its report on the communal conflict in West Kalimantan, conveys three arguments developed in accounting for the background of such conflict. Those arguments are the cultural argument, the marginalization argument, and the political manipulation argument. The cultural argument holds that cultural differences between the two disputing ethnic groups are the main factor. In connection with the conflict between Dayaks and Madurese, the habit to settle the dispute by way of bloody violence and fight is considered as the reason for the frequent occurrence of conflict between these two ethnic groups.

The marginalization argument embraced by many social observers considers the marginalization process of the local population, especially the Dayaks, as the consequence of various government's development policies that are very harmful to the native people. This triggers resistance and opposition of the local people

[21] In ethnical conflict cases in West Kalimantan, the Madura people had always become the party who suffered a lot, either in terms of the death toll, number of wounded people, or having to flee to other places.

[22] A review on this problem can be read in Teitelbaum and Winter (1998).

against anything coming from other regions. Migrants in this context are deemed as the representation of the outside party that must be got rid of. The political manipulation argument departs from the assessment that cultural and social–economic factors—although recognized as playing a role—are not strong enough to encourage a large-scale ethnical conflict such as the one in 1996–1997 between Dayaks and Madurese unless there is a third party manipulating such situation. The large number of arguments on the background of communal conflicts that always involved Madurese ethnic group in West Kalimantan indicates that the "ethnicity" inherent in the Madurese is only a small part of the communal conflict that occurred.

Perceived Threat and Perceived Advantage

Viewed from the political demographic perspective, border areas between West Kalimantan and Sarawak have a significant meaning and role. Those border areas are significant not only because of the objective reality as a territory bordering another country, but especially because of the subjective perception of various elements related to border areas. Viewed from the political demographic perspective, it is the subjective perception, especially of the incumbent elite that evidently plays a larger role in determining various policies and programs related to border areas. West Kalimantan border areas did not receive any attention from the central government until Sukarno and military commanders saw the threat posed by the establishment of Malaysia by the British. At that time, the border was perceived as a vulnerable area that needed to be immediately strengthened to become the basis of confrontation and at the same time to obstruct the infiltration from Malaysia. Perceived threat departs from the military elite's considerations placing state security and territorial integrity as a target and not as facilities to achieve the people's welfare.[23] Perceived threat that must be continually observed, both from the infiltration of foreign parties and from the military authority in Jakarta, was a stance underlying various policies and programs in West Kalimantan, especially in border areas. It was in this perceived threat context that organization of the transmigration program from Java and village rearrangement program in the vicinity of border areas could be understood. The mobilization of Dayak ethnic group by the military to force the Chinese ethnic groups to stay away from border areas can also be viewed as the manifestation of perceived threat strongly held by the military. Although during the

[23] Institutionally, perceived threat owned by the military is realized in the form of a territorial command structure that is hierarchic in nature from the central level to the entire parts of Indonesia. This territorial command in its reality functions more as domestic security control and keeper, especially against various political activities, deemed to threaten the "national stability." In its further development, especially during the New Order era, the role of the military was not only limited to security and territorial integrity problem but also further entered the nonmilitary problem through the ABRI dual function concept. A comprehensive review on this topic can be read in Sebastian (1996).

New Order era, the transmigration program from Java, Bali, and Madura to rural areas in West Kalimantan was always covered up with demographic (reducing population density) and economic (spurring regional development) reasons, this actually departed from perceived threat toward the possibility of the rise of separatist movements based on the strengthening of the ethnic identity of the local people outside Java.

In its further development, especially following the end of the period of confrontation with Malaysia, the central government's perception of West Kalimantan was no longer based on the strength of perceived threat, but it had shifted to a perception placing West Kalimantan as a treasure that must be maximally exploited and utilized for the interests of various groups at the central level. This perception can be called perceived economic advantage, subsequently generating various policies and programs to exploit and capitalize on the wealth contained in West Kalimantan land. While the perceived threat came from the military elite, the perceived economic advantage came from the economists–technocrats serving as the architects of the New Order's economic development. In viewing space and territory, both tended to generalize and assume differences as *cateris paribus* and given. During the New Order era, the development ideology ideally combined the perspectives of the military and the economist–technocrats. Centralized, uniform, top-down, and economic growth-oriented economic development during the New Order era was complementary to the military structure and role as the state's agent controlling territorial integrity and security control.

Concerning border areas, the presence of a neighboring country, in this case Sarawak, was not only viewed as a perceived threat but also viewed as a part of an effort to increase "economic advantages." The initiative to construct roads connecting Pontianak and Kuching and the official appointment of Entikong as "gate points" between West Kalimantan and Sarawak are logical considering the strengthening perceived economic advantages since the New Order era. Viewed from the perspective of political demography, people of West Kalimantan, a province directly bordering on Sarawak, Malaysia, was always the "victim" of the ongoing "marginalization process." This was due to the strong perception of the central government considering West Kalimantan as a target of demographic–political engineering to cater to the political–economic interests of the rulers (civil and military) at the central level.

Conclusion

Economic and political crises occurring in Indonesia since August 1997 have had quite an enormous impact on West Kalimantan. One of them has been the increase in illegal trading in border areas, especially in Entikong. There is an open opportunity to sell various Indonesian goods to buyers from Sarawak benefiting from the depreciation of Indonesian rupiah against Malaysian ringgit. Another impact is the increased number of Indonesians trying their luck by working in Sarawak, either legally or illegally. West Kalimantan has become an entry point for migrant workers from various places in Indonesia intending to enter Sarawak. Political changes occurring following the fall of Suharto in May 1998 have been the political decentralization and the granting of broader autonomy to regions.

For West Kalimantan, as well as other provinces in Kalimantan, the open opportunity for the local people to articulate their political and economic aspirations is expected to change the central government's perception in viewing problems in border areas. Border areas should not be merely viewed from the aspect of the central government's interests, which emphasize the security aspect. Serious attention must be given to promote the people's welfare. Without a planned welfare improvement program, people living in border areas will easily view their neighboring country as a reference in various aspects of life, especially in seeking job opportunities. In addition to that, the improved opportunity to articulate locally based demands is assumed to cause larger impacts on political and economic dynamics in the region, which may be marked by increased interethnic tensions.

Part III
Population Mobility Across Borders

AS WE ENTERED the twenty-first century, movements of peoples across national state boundaries are becoming global phenomena. The ending of the cold war and the breaking up of Eastern bloc countries have coincided with the influx of people into the West. The dramatic collapse of the Berlin Wall epitomized the unification of the German state, not from any military onslaught but from the exodus of its population. While social conflicts arising from the collapsing political systems are certainly behind the movement of the people, the economic tensions originating from the restructuring global capitalism are also forcing labor mobility on a world scale. In the global economic context, Southeast Asia has been regarded as the dynamic regions in the last three decades. In these regions, however, the economic development gap which coincides with demographic structural transformation has created labor market tensions within and between countries which inevitably have become push and pull factors behind the movements of people crossing national borders.

Studies on the political dimensions of international migration, particularly in the Southeast Asian context, are relatively few compared to the abundance of studies on the economic side. Adopting Myron Weiner's approach which places international migration within the Security/Stability Framework, the papers in this part will emphasize significant political dimensions in Indonesia and the important role of regional political grouping, such as ASEAN. In this context, special attention will be given to the political implications of Indonesian labor movements to neighboring countries such as Malaysia and Singapore. The lack of protection measures on Indonesian labor abroad as illustrated by the frequent violence experienced by the Indonesian workers cannot be comprehended without a clear understanding of the bilateral political relationships between the sending and receiving countries. Furthermore, the domestic political systems that excluding the working class in the political system has been very crucial in the foreign employment policy formulation and its implementation.

The three chapters in this part will emphasize the movement of people across the borders. Firstly, the international migration in Indonesia will be analyzed by placing the country within the Southeast Asian contexts. Secondly, analysis will be given to the politics of contestation between state and non-state actors in the politics of regulating overseas migrant workers in Indonesia. In the third chapter the conventional view of the division between internal and international is challenged theoretically and empirically, by using the Indonesian case studies.

Chapter 7
Indonesia and Its Neighbors: Political Dimension of International Migration

This chapter discusses the political dimensions of international migration, emphasizing Indonesia and its immediate neighboring countries. This discussion will use the security/stability framework (SSF) perspective as expounded by Myron Weiner. Relying on this perspective, the paper will endeavor to identify issues relevant to international migration, particularly those experienced by Indonesia in relation to its neighbors. It is important to note that Indonesia is a country with the largest population in the region and occupies a very strategic economic and political position. Geopolitically, Indonesia is indispensable to the maintenance of regional stability and security, which is also experiencing rapid economic growth. The economic crisis that has plagued the country since mid-1997 will inevitably influence the economic and political balance in the region. The migration of people, a justifiable reaction to Indonesia's declining economic and political situation, is a new social phenomenon that requires immediate attention. In the context of rapid political and economic changes, population mobility between countries is an important issue inseparable from the changes that have taken place. It is hoped that this chapter will be able to stimulate the thinking of the reader in order to discuss issues of international migration in Indonesia that, with no shadow of a doubt, will become increasingly important in the near future.

International Migration Framework: SSF vs. IPEF

According to Myron Weiner, economic experts have thus far discussed many issues concerning economic differences or inequalities between countries that influence migration. Political scientists have also discussed conflicts that arise in a country leading to a wave of political refugees leaving the country. However, very few experts have given the necessary attention to how international migration creates a conflict within or between countries, placing international migration as an

R. Tirtosudarmo, *From Colonization to Nation-State*,
https://doi.org/10.1007/978-981-16-6437-3_7

independent variable.[1] According to Weiner, a discussion like this is very important in understanding why a country and its citizens always have a negative attitude toward international migration, despite being aware of the economic benefits brought about by this kind of migration.

Weiner further explains what is meant by the SSF, which he carefully differentiates from what has long been known as the international political–economic framework (IPEF). IPEF explains international migration mainly by focusing global inequality, the economic relationships between the sending and the receiving countries. These relationships include the movement or shift of capital and technology, the role of the transnational institutions, and structural changes in the labor market closely related to the international division of labor. SSF focuses on national policies concerning international migration, created to address its impact on internal political stability and international security. SSF emphasizes the understanding of political changes within a country as the primary determinant of international and internal migration, including the wave of refugees both as a cause and an effect of international conflicts.

IPEF and SSF hold visible similarities. First, both shifted from an approach that sees migration because of individual decision making to an approach that looks at migration in the context of broader social, political, and economic changes. Second, both use an interactive framework that emphasizes the relationship between the process of migration and global change. Lastly, both also pay attention to the behavior of a country and the importance of its national borders. However, compared to IPEF, SSF stresses the importance of decisions taken by a country. In contrast, the role of a country is always treated lightly by IPEF, which considers global strength as the main determinant.

Both concepts also have different arguments on the aspects of international migration. They raise different questions, advance different explanations about international mobility, and provide different concepts of analysis. Although both reinforce or supplement each other, the lines of argument advanced are seldom the same. IPEF, for example, may use an analysis that looks at population mobility from a poor to a rich country as something of mutual benefit (the poor benefits from remittances, whereas the rich benefits from cheap labor it lacks). SSF, on the other hand, may see the same population mobility as a political consequence caused by changes in ethnic composition in the receiving country, which ignites conflict between migrants and the local community and further creates friction in the relationship between the two countries. This IPEF approach can also conclude that migration leads to a brain drain from the sending country and worsens the unemployment and housing problem in the receiving country. From another perspective, SSF looks at migration as something that can bolster a country's and international

[1]This quotation about Myron Weiner thinking was got from his paper entitled: "Security, stability and international migration" (1992/1993). The SSF perspective from Weiner in literature about "population and conflict" can also be found in a variant, which is slightly different for instance the thinking of Choucri (1984), Homer-Dixon (1994a, b), and Kaplan (1994). For the case in Indonesia, see Tirtosudarmo (1995, 1997).

peace, noting that migrants from ethnic minorities who are not socially accepted in their countries of origin can then be welcome in the countries to where they migrate. A cost–benefit analysis can therefore lead to a different evaluation and policies depending on the framework it applies.

According to an expounded analysis based on IPEF, international migration often connotes two essential political elements. First, international migration usually occurs because it is supported or discouraged by the government for reasons that do not have linkage to economic realms. For example, according to Weiner, the international migration in Africa and South Asia has little or no relationship with global changes or political–economic changes in the two regions. Secondly, suppose that economic factor is the cause of migration, it is the government that determines whether these people should be allowed to leave their country of origin and it is ultimately the receiving country that decides whether to accept or reject these migrants. Likewise, government policy is not always based on economic considerations.

Furthermore, governments have variations in their ability to control migrants entering the country. A country may be able to use military power to defend itself from foreign aggression but may not have the power to defend itself from migrants who enter the country looking for job opportunities illegally. A country that endeavors to control illegal migrants may not have the ability to stop them, but will look at it as a threat to sovereignty.[2] For this reason, therefore, understanding the political dimensions of international migration is vital in analyzing the phenomenon of international migration.

Indonesia, the Sleeping Giant of Southeast Asia

Today, Indonesia has a population of more than 200 million people, the fourth largest in the world after China, India, and the USA. It occupies a unique geopolitical position because it bridges the Asian countries, Australia, and the Pacific. Therefore, it becomes seemingly important to understand the population dimensions of Indonesia compared to other neighboring countries, such as Malaysia (18 million people), Singapore (three million people), and Australia (17 million people). Indonesia is undoubtedly a giant in Southeast Asia and its neighborhood if we gauge the country from the size of its population. Albeit ancient, theories that advance how large population influences a country's political power and its relationship with other countries are still very much adhered to.[3] The political implications of Indonesia's

[2] A study on Malaysia clearly shows how emigration of unskilled laborers from Indonesia, Philippines, Pakistan, and Bangladesh has an implication on national stability and security because of the high sensitivity of the racial balance in these countries (see Abdullah 1997).

[3] Finkle and Mcintosh believe that the fear of academicians of this long about a large population size and a high rate of population growth led to political aggression has never been proved right. According to them, there has been new confidence that a high rate of pop growth in the third world

population compared to its neighbors have not been examined adequately in international relations and discussions among academicians. It is, therefore, not easy to evaluate the impacts.[4]

In this context, the position of Indonesia in the Association of Southeast Asian Nations (ASEAN) is crucial and strategic, particularly in creating political stability and regional security. The birth of the Association of Southeast Asian Nations, which was a result of American influence, cannot be isolated from the Cold War conflict between the Communist Block and the Western Block. The fear of the Western Block, especially the USA, about the political orientation of Indonesia, moreover with the US defeat in its effort to subdue South Vietnam from the offensive of the Vietcong, was decisive. The fall of Sukarno and the collapse of Communism (Indonesian Communist Party) in 1965 provided the momentum for the West to influence the political re-orientation of Indonesia. At the same time, the confrontation between Indonesia and Malaysia was settled and directly led to the creation of ASEAN as a regional organization aiming to increase economic cooperation while advancing Indonesian political potential as a regional superpower in Southeast Asia.

Before the advent of colonialism, the geographical range of emigration from what we know today as Indonesia was pervasive: stretching as far as the island of Madagascar in the Indian Ocean to the South Pacific Islands. At that time, there were no rigid national state borders as is the case today. The Dutch colonization of the archipelago, which attained its shape in 1910, signified the creation of distinct territorial borders of the country, which later came to be called Indonesia.[5] Except for Thailand, the territorial borders for countries in the Asia Pacific region were inherited from the colonial powers. This was true of Malaysia (Britain), Brunei (Britain), Indonesia (Holland), Vietnam, Cambodia and Laos (France), the Philippines (Spain and America), Papua New Guinea (Britain and Australia), and other

is an obstacle to development "The old politics of population has been replaced by the politics of family planning," says Finkle and Mcintosh (1994).

[4] In a seminar about Malaysia at ANU (November 1997), Australia, Chandra Musaffar, a Malaysian activist and intellectual criticizes Indonesian labor policy as not having a clear structure and aim, appears not acting as was expected from a big country like Indonesia. In handling this critic, Hadi Soesatro, an economist from CSIS, attempted to defend himself by pointing out that Indonesia indeed no longer needs to "make noise" anymore just so people are afraid of Indonesia "bigness."

[5] Ricklefs (1981), interestingly portrays a particularly important episode from Indonesian history. According to him, ten (10) years after Diponegoro War (1825–1830), the Dutch started focusing attention to the Islands outside Java. Since 1840s, the Dutch extended their colonial authority outside the island of Java. There are two main reasons underlying the expansion of Dutch authority: (1) the Dutch felt that there was a need to have control over areas outside Java. They feared that the region might support an opposition movement against areas, which have been brought under the Dutch colonial rule; (2) the Dutch need to authenticate their territorial claim to other European countries who were at that time also competing for areas to colonize. In 1910, the borders of the country today called Indonesia were already under firm control of the Dutch colonial army with countless number of victims, the destruction of social ties, and humanitarian values and independence.

small countries in the South Pacific Islands. The colonization and decolonization process, which led to the creation of nation-states, has eventually transformed population migration across territorial borders into an unprecedented international problem.

Social, economic, and political pressures that arose following the migration across postcolonial nation-states in the Asia Pacific region, particularly Southeast Asia, are not only based solely on the developments that occurred over the last two to three decades, but also on the historical legacy of both the colonial and the precolonial era. For example, the problem of Indonesian workers in Malaysia, which is one of the main problems in the bilateral relationship between the two countries, is not simply a contemporary economic problem but also a historical problem of the migration process of the Malay people even before the advent of colonialism. Likewise, the people from Sangir Talaud in Mindanao in Southern Philippines and the people of Irian Jaya in Papua New Guinea have a long history movements that violate the postcolonial-made nation-state identities.[6]

As a country comprised of islands, the Indonesian archipelago is best characterized by its maritime culture and the constant movement of its population. This culture of wandering is deeply entrenched among ethnic groups outside Java, such as the Bugis Makassar, Buton, Madura, and Banjar.[7] The Minangkabau tribe of Western Sumatra, although known as wanderers, is not a maritime tribe like the Bugis Makassar and Madura. Besides the habit of wandering, which also takes the form of voluntary migration inside Indonesia, there has been planned migration for a long time. This is clear from the example of contract laborers or transmigrants. The Dutch recruited people from the Island of Java during the colonial period to work as plantation workers in the coastal areas of East Sumatra, Suriname in Central America, in New Caledonia in the south Pacific, and in Vietnam.[8] Other than that, transmigration, which continues from the colonial days, has led people from Java to spread to various islands to open up new lands as farmers.[9]

In Indonesia, political migration—the kind of migration intended to support the political interests of the ruling elite, used directly or indirectly in its literary meaning—has been ongoing right from the colonial era to the present day.[10] Generally, two objectives are intended to be achieved through political migration. First, it is a form of a systematic effort by the state to transfer people or encourage them to move from one place to another. The purpose is to create political security and social order in a region that is experiencing social difficulties or seen as having a potential

[6]There are about ten thousand (10,000) Indonesians in Mindanao, and some of them have lived there for three generations (*Kompas*, February 26, 1998).

[7]See Naim (1979).

[8]See Breman (1997), Suparlan (1995), Adam (1994).

[9]See Chap. 1, "From emigration to transmigration" in this book.

[10]See Tirtosudarmo, "The state's response to migration" in *Sojourn* (1998).

occurrence of insecurity or social unrest.[11] Colonization and transmigration are examples of political migration aimed at achieving the first objective. The ruling elite employed it to reduce the possibilities of social unrest in Java and remove possibilities of political instability in various places outside Java, such as Aceh, Irian Jaya, and East Timor, as they were seen as a threat to national integration.[12] The second objective of political migration, both during the colonial and postindependence era, was to create centers of economic growth outside Java, which was important to owners of capital protected by the ruling elite since they represented their interests.

Political migration in Indonesia, both in the past and the present, has a very distinctive character due to its inward-looking perspective. Several factors help explain the aspects underlying the creation of this inward-looking political migration. First, there is an excessive fear within the ruling elite regarding the social, political, and economic impacts of overpopulation in the island of Java. Second, Indonesia's geographical makeup is composed of thousands of islands, which further brings about fear among the ruling elite about the possible movement of separation of the local areas from the center.

The strong geographical groupings of people based on race, tribe, language, and religion have motivated the ruling elite to emphasize "unity" and national integration.[13] The two characteristics mentioned above cannot be isolated from internal politics dominated by the military, particularly the infantry, throughout the New Order regime. In the military's strategic framework, the spatial territorial factor is vital. Therefore, the program of population redistribution becomes instrumental in territorial military strategies.[14]

Indonesia, in the regional context of Asia and the Pacific region, and particularly Southeast Asia, is indeed a sleeping giant. Its large population, size, and the political dynamics of its migration, which is inward-looking in nature, are significant for the national response to international migration. The impact of ASEAN, which is indirectly a creation of the politics of containment by the West—mainly by the USA—on Indonesia's foreign, also contributes to the indecisive political attitude of Indonesia toward the inhumane treatment suffered by Indonesian migrant workers

---

[11] Social unrest among Javanese farmers as a reaction to colonial political-economic pressure between 1900 and 1920s, took place in Tangerang, Pamanukan, Sukabumi, Ciasem, Kuningan, Pekalongan, Gombong, Semarang, Mojokerto, Sidoarjo, Kediri, and Jember (see National Records RI 1981; see also Kartodirdjo 1973). In the beginning of 1950s, there was a transmigration of ex-freedom fighters and ex-soldiers from areas around Jakarta to Lampung. Among others, high concentration of these ex-guerrillas can become a source of trouble and instability, noting that during that time, a rebellion of DI/TII also occurred in West Java (See Heeren 1979).

[12] See Chap. 1 "From emigration to transmigration" in this book.

[13] There is a question about tribe in postindependence pop census based on the fear that a strong tribal identity may weaken the feeling for national unity (see Naim 1979).

[14] General Benny Moerdani, as the army chief of staff, while opening regular course on LEMHANAS, pointed out explicitly that transmigration is a program, which has got direct relationship with the defense and security system (*Kompas,* March 8, 1985).

abroad. The weakness of Indonesia toward the capital punishment imposed upon a female Indonesian worker in Saudi Arabia last year shows the lack of a clear protective mechanism in policies concerned with the sending of Indonesian workers abroad. Indonesia's obscurity in issues concerned with its migrant workers reflects a lack of political confidence in the country's foreign policy during the New Order. Big as it is, Indonesia has consistently failed in its international relations.

The financial crisis, which has become a full-blown economic crisis, has seriously accelerated social unrest and gave rise to various forms of violence in numerous cities in Java and outside the island of Java. There is a genuine fear of a migration exodus from Indonesia to various neighboring countries due to people seeking a safety valve to this crisis. Malaysia, Singapore, and Australia have already expressed fear in case this exodus happens.[15] There are at least three categories of migrants that are expected to leave the country because of this economic crisis. The first group consists of workers who are expected to enter the neighboring countries, especially Malaysia, illegally. Ethnic Chinese who own sufficient capital to migrate to other countries, especially Singapore, Canada, and Australia, conjures up the second group. The third group is expatriates who see Indonesia as a country that can no longer provide job guarantees and security. Speculations assume that the Indonesian government will not succeed in controlling population migration, given the severity of the economic crisis. As we approach the twenty-first century, the phenomenon of Indonesian migrant workers is taking a new shape. As we approach the twenty-first century, the phenomenon of Indonesian migrant workers is taking a new shape.

Migrants are no longer recruited to work as plantation laborers in East Sumatra or Surinam, but as unskilled laborers in Malaysia and housemaids in Singapore, Hong Kong, and Saudi Arabia. There has been a dramatic change in the fate of Indonesian migrant workers throughout this century. Additionally, historical reality suggests that the phenomenon of migrant workers cannot be isolated from the globalization process, both in the past and the present. Globalization, which is signified by the movement of people and capital across national state borders, is not a new phenomenon. Since the sixteenth century, the Dutch, Portuguese, Spanish, English, and French were the dominant players in the global arena. They competed to control the seas and oceans across continental borders, control other countries, and create colonies that subsequently came under their control. In these colonies, plantations and mines were created using labor recruited from places that were densely populated. The island of Java and mainland China became the most lucrative sources of cheap labor that could be sent to any place whenever there was a need to open plantations and mines.

The government's role with respect to international migration, especially migrant workers in Indonesia, is not well documented. As far as its records are concerned, the state's role is seen in the various forms of regulations and laws passed to respond to

[15] See press over the same issue. For example, see *Jakarta Post* (February 23, 1998), "KL steps up patrol to check immigrant influx" and *Jakarta Post* (February 26, 1998), "Possible exodus from RI worries region."

the need to control or organize the international migrant workers. Several ordinances were put in place by the Dutch colonial government toward the end of the nineteenth century to organize the workers for plantations in Sumatra. These laborers came not only from the island of Java but also from China. During the Japanese occupation (1942–1944), Indonesian workers were forcefully sent to Singapore and Thailand for the construction project of railroads and airports (Kurosawa 1993, 145, quoted from Effendi 1997).

Information about international migration in the postindependence period is still scarce. The relatively low volume of international migration in the postindependence period is perhaps one reason why information and studies on international migration are still limited. The government's response toward international migration in the postindependence period is not widely documented. However, various studies on migration flow indicate that the migration of people from Indonesia to countries such as Saudi Arabia, Malaysia, and Singapore started way back, even before the colonial era, and has continued after independence. It is only after the New Order regime came to power in the late 1970s that the Indonesian government explicitly started giving due attention to the increasing international labor migration.[16] In 1988, the former Minister of Labor Cosmas Batubara, who was inspired by the reality that the Indonesian workers' volume of international migration was increasing, decided to pass a Ministerial Decree No. 5, setting out the procedure for sending migrant workers abroad. Coinciding with the sending of workers to Saudi Arabia in the same year, the Minister passed another Decree No. 1307 about the technical requirements of Indonesian workers migrating to Saudi Arabia. As such, it can be said that the government had begun to give a direct response to the sending of migrant workers abroad from this moment. In 1994, the Minister of Labor Abdul Latief formed PT. Bijak, which helped organize the sending of skilled Indonesian workers to Malaysia.

The state's perception of international migrant workers, explicitly included in the GBHN Repelita V, primarily concerns the need to increase national foreign exchange earnings and reduce unemployment within the country, which, if left unchecked, might prove a threat to national stability. The problem is that globalization has its paradox. As the world becomes more open, the gap in income between the rich and the poor countries becomes wider. Free trade and free market mechanisms, which will inevitably be a reality for the future generation, will create a world that is more open and competitive, and hence, the strong will no longer be fair to the weak and the poor. That is the kind of dilemma faced by Indonesia, a country that is poor and has a vast population. The prevailing economic crisis that results from the fall of the value of the rupiah vis-à-vis the dollar has made it even more difficult for Indonesia to withstand these global forces.

[16] Since the beginning of the New Order, migrant workers going abroad have been quantified as follows: Pelita I (1969–1974) 5624, Pelita II (1974–1979) 17,052, Pelita III (1979–1984) 56,410, Pelita IV (1984–1989) 292,362, and Pelita V (1989–1994) 558,296 (quoted from Effendi 1997, 7).

For the last 6 months, we have witnessed thousands of our workers repatriated from Saudi Arabia and Malaysia. The country has been forced to extend a helping hand to these repatriated workers. Interestingly, the National Strategic Institute, which for a long time has served as an instrument of security, is being used as an institution assigned to handle and control the process of the repatriation. The state institutions that should legitimately handle labor problem, such as the Ministry of Labor and the Ministry of Foreign Affairs, have been subordinated during this crisis. It is clear, therefore, that the government's perception of the problem of migrant workers is based more on political and security considerations than economic prosperity.

The policy of the government toward its population, particularly its migrant population, has not changed much since colonial times. During the colonial era, postindependence era, to the New Order period, elite entrepreneurs have viewed labor as both a threat and a force to exploit. The question becomes how to reduce this threat and at the same time exploit these workers. Governments have, therefore, played a role as a force that uses but does not protect its workers. The plight of thousands of Indonesian migrant workers in Saudi Arabia and Malaysia is evident, and yet the government has not created an institution that protects international migrant workers. The establishment of the labor bill, which was forced to be completed within a compressed timeframe, and the "labor social insurance scandal" that followed the bill reflected the government's efforts to control labor activities. From the political–economic context, the phenomenon of migrant workers is evidence that the national development strategy in place does not target community welfare. Migrant workers, like any other workers within the country, are seen as a cheap source of labor in the process of increasing national economic growth. Politicians structurally and systematically weaken the position of workers since they look at workers as a threat to national stability, which is seen as a precondition for economic growth.

From a short-term perspective, a combination of a return of thousands of migrant workers from abroad and the country's economic crisis is likely to worsen the unemployment situation. The government's plan now is to create projects to absorb the unemployed, a step that should have been taken some time back. The question arising here is whether this crash program will be able to solve the existing problem. If it cannot, then there is a high possibility that a wave of illegal migrant workers will move to neighboring countries, especially Malaysia. It is expected that this movement will increase, although the economic situation in Malaysia will also be difficult because of the current economic crisis they are facing. From a long-term perspective, international migration right from the beginning is integrated into a social, economic, and political agenda of transformation, which must now be carried out. The time has come for the government to determine the individual position of migrant workers: its citizens who have economic needs and deserve respect as human beings.

Political Instability and International Migration

As a poor country with a large population, Indonesia is more of a migrant source than a destination. Looking at the SSF as expounded by Weiner, it is clear that domestic political conditions are significant determinant factors of how a country responds to international migration, which will continue to increase in the future. Inward-looking in nature and displaying a lack of respect for workers, the Indonesian politics of migration is an internal obstacle that will continue to determine the nature of the government's response to international migration. At the regional and global level, Indonesia's indecisive foreign policy will hamper those who wish to create a mechanism of protection for Indonesians working abroad.

At this time, the impact of the economic crisis on illegal migration to neighboring countries is a challenge that needs to be addressed. Migration is expected to increase as a reaction to the current crisis.[17] Within a short time, the increase in illegal migrants from Indonesia to neighboring countries like Malaysia, Singapore, and Brunei will have a pressing political impact on their relationships. Today, in Malaysia and Singapore, there are already many Indonesian migrants working there. This has become a very explosive internal political issue in Malaysia, where the number of illegal workers is estimated to be around one million.[18] The increase in illegal migrants from Indonesia due to the current economic crisis will certainly increase the internal political friction in Malaysia. Although thought to be having a small number of illegal migrants at present, Singapore will receive more due to the current crisis experienced in Indonesia. A political analyst from the Institute of Southeast Asian Studies has expressed fear about the far-reaching effects of the economic crisis in Indonesia, "If Indonesia's growth rate plummets, there could also be an accelerated migration of people to neighboring countries."[19] The economic crisis that occurred due to the fall in the value of the domestic currency vis-à-vis the US dollar can deteriorate into a political crisis. The People's Consultative Assembly, which is sure to reelect Soeharto as the president for another term (seventh), has major political implications for migrant workers who expect quick solutions to the

---

[17] The fear of possible increase in illegal migration to neighboring countries has indeed become a reality. For instance, over the last 2 weeks, the government of Malaysia arrested about 3000 illegal migrants attempting to enter and settle in various parts of the country. This figure shows an overly dramatic increase compared to less than 9000 throughout 1997. Besides, 26 Indonesians were arrested while attempting to enter Singapore illegally over the last 1 week alone (*Jakarta Post*, March 1, 1998). At around the same time, about 4000 intending illegal migrants were in various parts of Riau where they failed to enter Malaysia because of increased security and patrols along the Malaysian boarder (*Kompas,* March 1, 1998). Besides that, 40,000 illegal migrants were feared dead when a boat carrying 72,000 Indonesia captured (*Kompas*, Wednesday, February 4, 1998).

[18] See *Jakarta Post*, December 17, 1995, "foreign labor a major issue in Malaysia." See also Abdullah's study in 1997.

[19] See *Jakarta Post,* July 20, 1997. The political and economic risk consultancy, for example, pointed out the increase in the rating risk from 2.96 in December 1997 to 3.16 in January 1998, and speculating of the prospects of "population exodus" from Indonesia (*Jakarta Post*, February 26, 1998).

current crisis. If President Soeharto and the new cabinet fail to control the crisis after the ongoing Consultative Assembly, the government will appear to have lost its legitimacy and credibility, and it may be impossible to avoid a serious political crisis. A simultaneous political and economic crisis creates a combination of "push factors," which will force even greater numbers of migrants abroad than is the case today (footnote 19).[20]

Assuming that mass media reports and estimates about Indonesian migrants to Malaysia, Singapore, and Australia, are accurate; it means that no efforts are in place to monitor population mobility across the borders. The borders concerned connect Eastern and Western Kalimantan with Eastern Malaysia (Sarawak, Sabah, and also Brunei), Irian Jaya and Papua New Guinea, and North Sulawesi with Mindanao in the Southern Philippines.[21] There is also the potential for illegal population mobility from East Nusa Tenggara and East Timor to northern Australia. Ethnic and racial similarities represent a very influential social–cultural factor in creating migration networks across countries.

In a crisis like the current one, it is expected that ethnic and racial ties between Indonesians and people in the destination countries will play a significant role in international migration. The creation of domestic, economic, and political stability within a short time, as per Weiner's SSF perspective, is imperative in preventing illegal international migration to neighboring countries. In the context of political change, reformation in the procedure of sending Indonesian migrant workers abroad by the government should be carried out in a planned manner based on the principles of transparency and accountability in making decisions and public policies. As we face challenges for the future when the flow of capital, goods, and people between countries intensifies, is there a possibility that the current reactive response and role

[20]The Australian Embassy in December 1997 allowed 132 business migrants from Indonesia, and this figure portrays the dramatic increase than the months before which registered only between 30 and 40 applicants (*Jakarta Post,* February 26, 1998).

[21]Population migration of tribes in Kalimantan and Irian Jaya across national borders is something usual and has been a common practice over an exceptionally long time. This cross-border migration will experience a drastic increase when there is social unrest and natural disasters. The flow of refugees from Irian Jaya to Papua New Guinea, which occurred several times in the 1980s, was a reaction on threats and instability in places where they were living. At the time when there was a conflict between the Dayak and Madura tribes in Sanggau Ledo, West Kalimantan in December 1996 and February 1997, there occurred unprecedented migration to Sarawak. Besides social unrest, political instability and natural disasters like drought and famine are push factors to population migration, which always takes place. Migration as a consequence of environment changes or resource scarcity is a new issue, which has just been addressed as a social science (see Homer-Dixon 1991; Swain 1996). In 1997, hundreds of people in the remote highlands of Irian Jaya died of famine caused by draught and epidemics, which invaded that area. At this time, in the remote areas of Kutai, it is said that about 8000 people are facing a threat of starvation because of draught (*Kompas,* March 2, 1998). If there is an opportunity for people in these places of Irian Jaya or Kalimantan who are currently suffering, these disasters' migration to neighboring countries, which have better environmental and economic prospects, cannot be resisted. "Environmental migration" (a terminology coined by Swain 1994) has got a political impact for both Indonesia and its neighbors as places of migration destination.

of government continues? There is a need for a more proactive government attitude in handling the problem of migrant workers in the future. This kind of attitude assumes a change in perception to the existing one, which looks at workers as a threat and only a source of foreign exchange. This perception needs to change so that workers are looked upon as citizens who have rights that everyone should respect. There needs to be a feeling of humanity as a pillar for their legal protection to support policies for sending migrant workers abroad.

Chapter 8
The Politics of Regulating Overseas Migrant Workers

In Indonesia, the ruling elites generally overlook regulation policy on overseas migrant labor. The legacy of past experiences has been crucial in shaping the perception and responses, of the New Order's political and bureaucratic elite on the formation of institutions and regulation concerning overseas migrant labor. The high proportion of undocumented migrant workers and the absence of state protection measures indicate the lack of clear state policy on foreign employment. The business industry on overseas migrant labor is strongly influenced by nonstate actors, primarily the labor recruiting and supplying companies. The role of nongovernment and human rights organizations in articulating the interests of po-werless migrant workers was generally ignored during the New Order period. After the fall of Suharto, nongovernment organizations and migrant worker organizations have had more opportunity to pressure the state to establish a pro-overseas migrant labor regulation, among others, through advocating a new national law on migrant workers. In its further development, for example, the repulsive incidents of human rights and sexual abuses experienced by unskilled female overseas migrant laborers have even prompted women's movement groups to campaign against exporting migrant workers abroad. This chapter exposes the unfolding politics involved in the regulation of overseas migrant labor, particularly on the contestation between state and nonstate actors, in their attempt to influence the state's policy on the increasingly important phenomena of overseas migrant labor.

Migration in the State's Political Agenda

In Indonesia, internal migration has always been an important issue within the state's political agenda. This was mainly due to the perception that a range of the state's aims can be achieved through a migration policy, in the context of both colonial and postcolonial states. The state's policy on migration begun during the beginning of the twentieth century following the intense development within the colonial elites

R. Tirtosudarmo, *From Colonization to Nation-State*,
https://doi.org/10.1007/978-981-16-6437-3_8

that eventually culminated in the formulation of the so-called ethical policy. The ethical policy was mainly concerned with raising agricultural productivity and had not intended to introduce drastic changes in the colonial economic structure through large-scale industrialization. As Legge (1964) has argued, the measures taken by "ethical" colonial governments to improve the living standards of the indigenous population were mere palliatives; they provided some alleviation of areas of hardship but did not achieve—and they did not attempt to achieve—any thoroughgoing technological changes. Many ethical reformers stressed the importance of large-scale industrialization, but until the worldwide economic depression of the 1930s, the various plans suggested were not implemented.

From a different perspective, Kartodirdjo (1973) perceived that sporadic movements of peasant unrest marked the rural history of Java in the nineteenth and early twentieth centuries. Many of these erupted in violent clashes with the colonial authorities. Protest movements and social unrest occurred between 1900 and 1920 in various places in rural Java, such as in Tangerang, Pamanukan, Sukabumi, Ciasem, Kuningan (West Java), Pekalongan, Gombong, Semarang (Central Java), Mojokerto, Sidoarjo, Kediri, and Jember (East Java).[1] Although a direct link between social unrest in rural Java and the initiation of the emigration policy cannot be established, it is not implausible to posit a cause–effect relationship. A major complicating feature of agrarian unrest in Java has been its correlation with social change in general and with the colonial impact. Kartodirdjo (1973) strongly argued that the social movements had their background in the rapid penetration of a colonial economy whose impact on rural Java reached a climax during the nineteenth century. The colonial rulers introduced a new legal and social relationship covering agrarian and labor matters. Excessive demands for compulsory services from the population and the levying of new taxes exacerbated popular discontent. Turning over some land to sugar cultivation and the exaction of a compulsory contribution of paddy rice had a direct bearing on some of the instances of social unrest.

As one of the three ethical policy objectives, the Dutch colonial migration policy (emigratie) was therefore formulated and maintained because of a combination of three main factors. First were the political changes in the Netherlands, which allowed the Calvinist–Catholic coalition to come to power in 1901. The outstanding feature of the policy outcomes from this new coalition was the official abandonment of the goal of economic exploitation and the introduction of direct intervention in the economic sphere to improve the conditions of the indigenous population. The second factor was economic opportunities, particularly as seen by the Dutch capitalists after the entire archipelago was successfully brought under effective colonial control. The vast land areas on the Outer Islands attracted private companies looking to establish plantations. Given the scarcity of labor, Javanese were recruited as cheap labor for new economic activities. The third factor was social unrest in many parts of rural Java, resulting from simultaneous economic exploitation and population

[1]These movements have been comprehensively documented by the National Archives of the Republic of Indonesia (Arsip Nasional Republik Indonesia) in 1981.

pressures. These, in turn, encouraged the colonial government to deal with social unrest by moving people to the Outer Islands. As such, the migration policy developed by the Dutch can be summed as a valuable instrument to serve the goals and interests of the state and its ruling elite. The relocation of people to ease the social and political tensions is a form of demographic engineering to serve the state's economic and security purposes.

About 10 years after independence, the government formulated for the first time its five-year development plan from 1956 **to** 1960 in which transmigration was described as an instrument for reducing population pressure in Java, supplying labor in sparsely populated provinces, supporting military strategy, and accelerating the process of assimilation (Hardjosudarmo 1965, 128–129). Regarding transmigration policy, the plan's most significant change was its explicit reference to strategic military purposes, making industrialization no longer the goal. This was apparently due to increased political unrest in some regions resulting from disappointment with central government leadership. The strategic role of transmigration was further emphasized in 1962 because of a change in the national constitution whereby President Sukarno proclaimed the so-called guided democracy system of government to replace the parliamentary democracy system that he considered to be a failure. The establishment of the autocratic guided democracy took place with the support of the central military leadership (Kuntjorojakti 1978, 138–139).

Nitisastro, from a somewhat nationalist stance, criticized the Dutch assumptions that had led to enclave politics (Heeren 1979, 17). According to Nitisastro, the typical Dutchman's conservative politics had hampered the assimilation process among ethnic communities in the resettlement areas. Furthermore, Nitisastro argued that enclave politics were obviously in contradiction with the primary goal of Indonesian independence, building one Indonesian nation. Assimilating the Javanese with the local population was expected to bolster unification. However, this goal proved to be more complicated than the government or nationalists like Nitisastro had anticipated. Wertheim (1959, 196), for example, noted from his observations in Lampung in 1956 that the assimilation of Javanese migrants resulted in neither a Sumatran nor a general Indonesian society, but rather a Javanese society modified by a Sumatran environment. This situation, according to Wertheim, led to the increasing resistance of *Sumatrans* to resettlement policies. Such resistance could seriously hamper further transmigration efforts, Wertheim argued, since the absorptive capacity of the Outer Islands is restricted not only by spatial and technical factors but also by social ones as well.[2]

At the beginning of the New Order, the real motive of President Suharto for the continuation of transmigration, however, is not difficult to identify. The explanation lies in the idea of harmony among the Javanese, which in the Indonesian political context can be translated into the concept of national unity and national integration,

[2]Wertheim's prediction about the probability of social conflict in Lampung because of rapid population growth and social tensions between migrants and local people occurred in the so-called Lampung Affairs, which broke out in February 1989.

as suggested by Koentjaraningrat, the doyen of Indonesian anthropologists, and strongly endorsed by the President and the military.[3] In this context, transmigration was perceived as an instrument to accelerate the process of national integration. Government policies to relocate people from overcrowded Java to other islands were also considered as an alternative to land reform.[4] Despite this, the legacy of colonial policies and the population policies of the Old Order, which were pronatalist and viewed the uneven distribution of population as the primary population problem, could not be easily eliminated from the thinking of the ruling elite.[5]

The conventional view among Indonesia's New Order economist–technocrats was that the rate of population growth is a significant determinant of the success of a country's economic development efforts. Compared with the problem of rapid population growth, which was accorded high priority at the beginning of the New Order, the problem of uneven population distribution between Java and Outer Islands had not been generally regarded by the economist–technocrats as an issue of urgency. Nitisastro (1970, 238), widely known as the architect of the New Order's economic development policies, strongly argued that what was needed to overcome the population problem in Indonesia was a massive development effort to create expanding employment opportunities accompanied by a rapid spread of fertility control. Nevertheless, uneven population distribution and the role of transmigration as a means of overcoming it had long been a focus of government thinking and became a significant policy within the national development plans during the New Order. The ruling elite's obsession with internal migration—a legacy of the colonial past—emphasized the inward-looking character of the state's perception of migration.[6]

Patterns and Characteristics of Overseas Migrant Workers

This section will show the magnitude of overseas migrant workers demographically, particularly concerning their migration pattern and characteristics. Table 8.1 clearly shows the rapid increase in the number of official migrant workers from less than 20,000 in the second plan to more than 800,000 in the sixth plan. The table also markedly shows the sharp increase in migrants to Saudi Arabia—which comprises

[3] "An Interview with Koentjaraningrat," Visser (1988).

[4] Land reform, which was aggressively promoted by the Indonesian Communist Party (PKI) prior to the 1965 abortive coup, was identified by the New Order as a communist policy.

[5] An illustration of how the elite, particularly the military, have persistently regarded transmigration as an important undertaking is shown by a request from General Suharto to the US Ambassador in Indonesia, Marshall Green, at their first meeting on May 29, 1966. In the meeting, Suharto asked for US$500 million in grants or soft loans to assist the transmigration program (Personal communication with Dr. Terry Hull, ANU, 1989).

[6] For more detail account on the continuation of the migration policy from the colonial to the postcolonial periods, see Tirtosudarmo (1997).

Table 8.1 Overseas migrant workers according to the destination country 1974–1999

Destination countries	2nd plan 1974–1979	3rd plan 1979–1984	4th plan 1984–1989	5th plan 1989–1994	6th plan 1994–1999)
Saudi Arabia	3817	55,976	223,573	268,858	267,191
Other Middle East Countries	1235	5349	3428	5145	16,071
Malaysia and Brunei	536	11,441	38,705	130,735	392,512
Singapore and Hong Kong	3729	6768	12,272	38,071	80,222
Korea, Taiwan, and Japan	451	920	573	6153	45,256
Other	7274	15,956	13,711	17,010	13,100
Total	17,042	96,410	292,262	465,972	814,352

Source: Hugo (1995, 279); unpublished official statistics from Indonesian Ministry of Manpower

Table 8.2 Indonesian migrant workers according to gender and destination country 1994–1997

	Male		Female		Total	
Destination countries	Number	%	Number	%	Number	%
Saudi Arabia	20,970	6.76	246,221	48.86	267,191	32.81
Other Middle East	795	0.26	15,283	3.03	16,078	1.97
Malaysia/Brunei	218,193	70.30	174,319	34.58	392,512	48.20
Singapore/Hong Kong	19,035	6.13	61,187	12.14	80,222	9.85
Korea/Taiwan/Japan	38,361	12.36	6895	1.37	45,256	5.56
Other	13,018	4.19	75	0.01	13,156	1.62
Total	310,372	100.00	503,980	100.00	814,352	100.00

Source: Various unpublished official statistics, Indonesian Ministry of Manpower

more than two-thirds of the total—in the fourth plan. Since the fourth plan, the number of migrant workers to Saudi Arabia has remained steady, whereas the number of official migrant workers to Malaysia and Brunei has constantly increased, tripled in number between the fifth and the sixth plan. In general, Saudi Arabia and Malaysia have become the major destination countries of overseas migrant workers from Indonesia. Several studies have shown that the decline of the proportion of migrant workers to Saudi Arabia during the last two five-year plan periods is partly due to the opening of new labor market opportunities in other Asian countries, particularly Malaysia, Brunei, Singapore, and Hong Kong. Many migrant workers, particularly in Malaysia, are employed in plantation and construction industries that are mostly dominated by males. However, a study carried out in a village of East Flores found an increasing tendency for women labor migrants from Indonesia to be deployed in domestic work in Malaysia (Raharto 1997, 3).

In the latest five-year development plan period, females account for more than 60% of all overseas contract workers recorded leaving Indonesia (Table 8.2). About 49% of those are still bound for Saudi Arabia, while the next highest percentages migrate to Malaysia. Only a small percentage of women migrant workers go to other Asian countries. The table also presents those female migrant workers participating

Table 8.3 Occupation and place of destination, 1997

Occupation	Middle East	Other Asia[a]
Housemaids	81.13	53.49
Drivers	13.21	0.00
Construction workers	3.77	4.65
Operator	0.00	6.98
Other	1.89	16.28
Total	100.00	100.00
N	53	42

Source: IPB, 1997: Table 2
[a] Note: Singapore, Hong Kong, Malaysia, Brunei, Taiwan, Korea, and Japan

in the labor market in Saudi Arabia outnumber males by 12 to 1 in the officially documented flows. These data are likely to be much higher if the undocumented or illegal flows were to be included. In fact, there are a considerable number of women going to Saudi Arabia for reasons other than work, most commonly for "*umroh*" pilgrimage, but who end up working.[7] As reported by the press, the numbers of migrant workers in Saudi Arabia are considerably greater than the official statistics show and account for 319,444 people. The number is lower than those who are in Malaysia (408,211), although it is much higher than those in Singapore (86,209) and South Korea (26,278).[8]

The female workers going overseas, including to Saudi Arabia, are generally in their prime age. Although statistical data provided by the Ministry of Manpower are not available, a study undertaken in one village in West Java Province identifies that the age of female migrant workers is much lower than that of males (Adi 1996, 228). The data of age selectivity show that the average age of females working in Saudi Arabia is 27 years, or about 10 years younger than male workers. Additionally, the age distribution of women migrant workers reveals that the majority are in the age group between 20 and 39 years.[9] Compared to males, fewer women are in the older age group, but more women are in the 15- to 19-year age group.

A survey conducted in West Java shows that most migrants are engaged in the unskilled and semi-skilled types of occupations (Table 8.3). Housemaids comprise more than 80% of respondents who work in Middle Eastern countries, whereas in other countries, only around 50% of the respondents work as housemaids. The data reflect that most women migrant workers are exposed to many risks and exploitation as their primary occupation is not covered by the labor laws of either the sending or

[7] A study conducted in 1986 (see Cremer 1988, 77) shows that half of the migrants cited making the pilgrimage to Mecca as one of their motives, and it was second only to getting a higher salary (61%).

[8] The Jakarta Post, November 1997.

[9] From interviews with women migrants and with resource persons, the ages of the migrants were usually upgraded to meet the criteria stipulated by the Saudi Arabian government. The minimum age requirement to work as a domestic worker is 20 years, but the girls who entered into this employment were mostly under 20 (see also Cremer 1988, 76).

receiving countries. Putting into account the working conditions of domestic workers, they are often subjected to abuse and violence in terms of economic, physical, and social/psychological well-being (Lim and Oishi 1996, 2; Cox 1997, 59; Shah and Menon 1997, 14–18). In promoting the export of women's labor, the government of Indonesia has taken steps to regulate the working requirements. The 1996 guidance for the placement of Indonesian labor to Saudi Arabia states that the prospective women applicants should apply for positions as overseas workers to the Local Office of the Manpower Department of the relevant regency with the complete required documents.[10]

Considering the complexity of the documents that the prospective migrant workers have to complete, it is often hard for them to fulfill all of the application and working requirements without help from someone else. It is frequently the case that, due to lack of education, many prospective women migrants get into difficulties when trying to understand what kind of letter/document is needed to complete the application processes, including the working requirements. Therefore, many women tend to use middlemen (usually recruiters) to take care of these matters, resulting in the possibility of falsification of documents. Document falsification is quite common during the recruitment process. This document falsification occurs at the village level, particularly in the process of obtaining letters—as it must be within the knowledge of the head of the village—and at a national level, usually in terms of passport falsification. At the village level, when completing the application form, the address and age of the prospective migrants are often faked with invented data, under the direction of a labor recruiting agencies' field workers. Such a situation contributes to fake data being present in all the official documents.[11]

Women workers, particularly those going to Saudi Arabia, should be treated differently, underscoring their distinctive contextual characteristics from the contexts of other overseas migrant workers. These distinct characteristics lead to the specific requirement for policymakers and other agencies concerned with women migrant workers to Saudi Arabia. The requirements particularly apply to labor recruiters and suppliers and NGOs working on the issues of migrant workers. There are four distinct contextual characteristics concerning women migrant workers to Saudi Arabia: (1) They are mostly unskilled and uneducated females engaged in informal domestic service types of employment; (2) the institutional structures lack precise regulations and enforcement that easily result in the situation where the women migrants become the most disadvantaged party in the whole process of the labor exporting industry, (3) Saudi Arabia, as a state and as a place for Indonesian

[10]The document comprises identity card, birth certificate or its equivalent, marriage certificate for those who are married, or a letter confirming they are widowed, divorced, or single within the knowledge of the head of the village, written approval from spouse or parents/foster parents, stamped under the knowledge of the head of the village, statement/certificate of skill/specialization in accordance with job requirements, and photographs and job enrollment card obtained from the Local Office of Manpower.

[11]Information on this part is derived from the author's study in 1997 (see Tirtosudarmo and Romdiati 1998).

women migrant domestic workers, has a distinct culture and legal system, (4) as major Muslim countries, Indonesia and Saudi Arabia have special bilateral ties, which to some extent influence the policy and management of Indonesian women migrant workers to Saudi Arabia.

The problems emanating from the increasing phenomena of female overseas migrant workers can generally be seen from macro- and micro-perspectives. The macro-perspective perceives the migrants as a part of broader demographic and socioeconomic changes at the national and international levels. Observations based on this macro-perspective approach resulted in suggestions of the possibility of changing or improving the existing structure of social and political instruments related to the overseas migrant workers. The micro-perspective that focuses on the behavioral dynamics of the women migrants draws attention to how the individual women migrants can be empowered to protect themselves from the often unknown and uncertain working environments they will encounter. However, to improve the working conditions and welfare of the women migrant workers, the two perspectives should not be separated.

State's Policy on Overseas Migrant Workers

As described earlier, both in the colonial and early postcolonial periods, the state gave little attention to issues of international migration. Apart from the fact that the number of voluntary international migrants was relatively small, it had not yet become a political–economic issue in international relations. Many studies, however, have shown that labor movements from Indonesia to other countries were practiced during the colonial period.[12] Following independence, particularly after the 1970s, the increasing international migration of unskilled laborers, mainly to Malaysia and Saudi Arabia, apparently had begun to attract the state's attention to international migration issues. It was not until the mid-1980s that the Indonesian government began to pay more attention to the increasing number of international migrant workers, particularly those going to the Middle Eastern countries. In 1981, a private but well-connected Indonesian Manpower Supply Association (Asosiasi Perusahaan Jasa Tenaga Kerja Indonesia, APJATI) was formed to regulate the flow of workers to the Middle East. APJATI was a consortium of labor recruiting agencies, officially licensed by the Ministry of Manpower to broker job contracts with employers abroad and arrange for the conveyance of workers. As the stream of workers to the Middle East increased, the state began to see overseas employment as a useful tool for its surplus labor problems, identifying it explicitly as such for the first time in the fourth five-year development plan (1984/1985–1988/1989). The plan

[12] See studies by Thompson (1947), Vredenbreght (1964), Suparlan (1995), Adam (1994), Breman (1997), and Miyazaki (1998).

set a target of 225,000 workers to be sent abroad to ease the difficulties of absorbing the rapidly growing labor force.

Around the mid-1980s, just as the government realized the perceived advantages of large-scale labor migration, the Indonesian press increasingly began to report cases of abuse against Indonesian women domestic workers in the Middle East, who constituted 87% of the total Indonesian workforce employed there. The state's response to migrant workers to the Middle East (primarily as domestic helpers) was instigated by the increasing press reports on the violence and sexual abuse experienced by the workers in Saudi Arabia. The state's response to the issue of the negative experiences of female migrant workers shows the lack of intention to create policy measures to protect the overseas migrant workers. Instead, the state's response was primarily aimed at saving the bilateral relations with the Saudi Arabian government and complained that the press reporting was biased toward the Saudi Arabian government. In 1985, a former Chief-Executive of the Operation to Restore Order and Security Command (*Komando Operasi Keamanan dan Ketertiban "Kopkamtib"*) and the acting Minister of Labor at the time, retired Admiral Sudomo, announced a ministerial decree stating that the migrant workers were not allowed to talk to the press about their experiences.

The Indonesian government and the public gave little attention to the labor movements in Malaysia. While in Malaysia the influx of Indonesian migrants had become a major political issue, it attracted little attention at home.[13] If there was any attention, it most likely came from the business groups behind the labor recruitment agencies. Since the early 1980s, the Malaysian government had begun to take serious steps in response to the public demand for tackling the problem of illegal migrant workers. In 1984, for example, the Malaysian government initiated an agreement with the Indonesian government (Medan Agreement) to regulate migrant workers, specifically in the plantation and domestic services sectors, the largest source of employment of illegal workers from Indonesia. This policy was a failure due, among other reasons, to the reluctance of the employers, both in the plantation and domestic sectors, to support the implementation of policy regulation. The flow of migration to Malaysia, both to the peninsula and Sabah and Sarawak in Kalimantan, has a long tradition as many Malaysians originate from Indonesia. Since the early 1970s, the rapid economic growth in Malaysia has escalated the welfare of its population resulting in the need for foreign labor for low-wage occupations. Therefore, the labor movement from Indonesia is a natural economic trend as Indonesia always has a labor surplus.

At the beginning of the fifth five-year development plan (1988/1989–1993/1994), a significant policy step was initiated in 1988 by the Minister of Manpower, Cosmas Batubara, who announced a Ministerial Decree No. 5, 1988, to regulate the overseas labor export process. At the same time, a special regulation (Ministerial Decree No. 1307) stipulating the Technical Guidance for the Recruitment of Labor to Saudi Arabia was also enacted. The beginning of the sixth five-year development plan

[13] On the political dimension of Indonesian illegal migration to Malaysia, see Chap. 8 in this book.

(1993/1994–1998/1999) was marked by an ambitious target of exporting 1.25 million migrant workers abroad. The new minister, Abdul Latief, a successful businessman, also decided to install a new Directorate of Export of Indonesian Workers replacing the former Biro of Foreign Employment office. A semi-government agency, PT BIJAK, was established to coordinate overseas labor recruitment. Another development during this period was the creation of the Indonesian Human Resources Foundation (Yayasan Sumber Daya Manusia), whose main task is to collect fees to provide insurance for the workers. However, the effectiveness of such an insurance policy is highly questionable as none of the government's assistance protected many migrants' legal problems. It was during Abdul Latief's period as labor minister that a new national labor law was proposed to the parliament. The draft of the labor law that the government proposed has aroused widespread criticism and public protest both by the intellectuals and academics, NGOs, and labor organizations. The draft law is generally perceived as representing the view of the state and its employers while disregarding the workers. The inward-looking character of the state's view on labor issues is also reflected by the absence of foreign employment guidelines in the new draft law. In general, the newly created labor law is strongly rejected by NGOs and labor activists as providing the state more power to control the increasing labor movement in Indonesia.[14]

Since August 1997, Indonesia has been hit by a severe economic crisis as the value of the Indonesian currency (rupiah) plummeted against the US dollar. More than ten million people are predicted to be unemployed in 1998. The currency crisis that hit the East and Southeast Asian countries has resulted in the deportation of Indonesian workers currently employed in Malaysia, Singapore, and South Korea. Malaysian officials, for example, have planned the deportation of around one million of its foreign workers, of whom Indonesians comprise the majority. President Suharto has instructed the Indonesian Minister of Labor to provide a labor-intensive program for thousands of workers who lost their jobs as the monetary crisis hit many construction works. Furthermore, as the 1998/1999 state budget has predicted a negative annual economic growth, the government will likely concentrate on the alarming social and political unrest resulting from the expected rapidly growing unemployment numbers and the decline in people's welfare standards. Mainly resulting from the lack of work permits, in November 1997, more than 35,000 undocumented women migrant workers were forced to return from Saudi Arabia, further aggravating the social and economic hardships in the rural areas as most of these migrants returned to their villages.[15]

Although the reported cases of human rights abuses are very high, the Indonesian government clearly had difficulties with the Saudi government in dealing with the problem of migrant workers Smith Alhadar, a researcher from the Society for Middle

[14] See Tirtosudarmo's account on labor movement and democracy in Indonesia (1998).

[15] For the difficulties and hardships experiencing by migrant workers during the economic crisis, see, among others, Chuzaifah (1999). On the macro-level analysis of the impact of the economic crisis, see Ananta et al. (1998).

East Studies with 5 years of working experience in a construction building company in Jeddah, argued that the weak position of the Indonesian government vis-à-vis the Saudi government resulted from the fact that Saudi has provided many donations to Islamic organizations in Indonesia. Other than that, Saudi Arabia is also influential among the International Islamic Organization through their control of the major supply of oil.[16] Furthermore, the control of two Islamic holy cities, Mecca and Madina, where Muslims from all over the world make their pilgrimages, provides the opportunity for the Arab Saudi government to decide on the number of people the Indonesian government shall be allowed to permit to make a pilgrimage. This is a critical issue for the Indonesian government since Indonesia is one the largest Muslim countries in the world. Naturally, the Saudi's power to control the numbers making a pilgrimage each year has considerably influenced the subordinate position of the Indonesian government to the Arab Saudi government. This subordinate position affects most areas that connect the two countries and certainly influences the Indonesian government's dealings in the problem of female workers in Saudi Arabia.[17]

The repercussions of the economic crisis and the repatriation of overseas migrant workers, particularly from Saudi Arabia and Malaysia, further halted the development of Indonesia's foreign employment policy and forced the state to focus on burgeoning unemployment. That is because both problems could lead to social disruption, which in turn could destabilize security and political order. Following the economic crises, the state naturally confronted the growing and intensifying pressure from the public—mainly students who are usually at the forefront of this type of political movement—to take clear steps toward solving the deepening economic crisis in the country. Suharto finally stepped down as the mounting political pressures could no longer be sustained. On May 21, 1988, Suharto handed over his presidency to Habibie, the Vice President. During his presidency, which only lasted for around one and half years, Habibie's government accommodated various demands from society, and among others, concerning labor. The workers should be free to organize and establish organizations and unions. The number of labor organizations was flourishing; many of them decided to form a political party and participate in the general election that was conducted in June 1999. The new government also revoked the newly granted national labor law that was highly criticized by the free labor unions and agreed to revise it accordingly.

However, the public euphoria that followed the collapse of Suharto's government had only little effect on the fate of overseas migrant workers. Besides the fact that the economic crisis generally affected many workers—noting that many factories had to close down, the issue of overseas migrant workers was obviously subordinated to the

---

[16]Nasib Solekah dan Inferioritas Indonesia vs. Saudi, *Republika*, November 23, 1997.

[17]The issue of *haj* quota and migrant workers in Arab Saudi has never been openly discussed presumably because of the sensitivity among the Muslims in Indonesia. Saparinah Sadli, the chairman of National Human Rights Commission for Woman, has revealed this matter during her speech in an open discussion forum organized by International Women Group on December 2000 in Leiden University, Netherlands.

wider political issues dominating public attention. The attempt to advocate the problems of overseas migrant workers by *Solidaritas Perempuan* (Women's solidarity), for example, by bringing the women migrant workers to meet with the Minister of Labor in the last Suharto's cabinet, Theo Sambuaga, a top Golkar leader, did not affect the policy. During the short period of Habibie's presidency, Fahmi Idris, another top leader of the Golkar Party, was assigned as the labor minister, which seems relatively active in accommodating the pressures from independent labor unions, such as *SBSI* (Serikat Buruh Sejahtera Indonesia or Indonesian Prosperous Labor Union), in ratifying several international labor conventions. However, the problem of overseas migrant workers remained beyond the scope of the mainstream labor issues. A slight move regarding the problem of migrant workers showed during this time by the relatively active new labor minister's deputy Din Syamsuddin—also representing the Golkar party—which was in charge of overseeing the overseas migrant workers.

As a scholar of Islam and activist of the Islam organization (*Muhammadiyah*), Din Syamsuddin appeared to have a strong interest in the fate of many women migrant workers who were suffering from various human rights problems in their host countries, particularly in Saudi Arabia. Din Syamsuddin showed his willingness to cooperate with the advocates of many migrant workers and seemed to be taking strong measures against the labor recruiter's organizations. In his term of office, tension increased between the government, the labor recruiter, and supplier organizations. The two were perceived as the source of problems regarding the lack of proper preparations in recruiting and exporting labor. During this time, there were apparent attempts to improve the policy on overseas migrant workers by the government decision to open a special terminal for overseas workers within the Sukarno–Hatta International Airport in September 1999. The opening ceremony of this special terminal was attended by President Habibie, at which (in his inaugural speech) he defended the need to improve the policy on sending migrant workers overseas. The Minister of Labor reported that an average of 600 workers were departing and 400 workers arriving in Sukarno–Hatta international airport every day.[18]

In December 1999, Abdurrahman Wahid was elected by the parliament as the president replacing Habibie. In the new cabinet, the Minister of Labor was from the Golkar Party, Bomer Pasaribu, a former leader of the backed government's labor unions (Serikat Pekerja Seluruh Indonesia—SPSI or Indonesian Federation of Labor). However, the reshuffling of the cabinet 8 months later, in August 2000, replaced both Pasaribu and Syamsuddin from the labor ministry. Alhilal Hamdi, a minister of Population and Transmigration, was assigned to merge the labor and transmigration issues into a single department. Soeramsihono, a senior member of staff at the Ministry of Labor, was also appointed to replace Din Syamsuddin. The appointment of Soeramsihono, a long career bureaucrat who had served in many posts within the Department of Labor, reflected the return of policy control of

[18] As reported by *Republika*, September 1, 1999.

overseas migrant workers from the politician to the bureaucrat.[19] This change will probably result in the slowing down of possible reform in regulating overseas migrant labor. By this time, the public campaign to stop sending women migrant workers to Saudi Arabia was heated. The Minister of Women's Empowerment, Khofifah, officially supported the idea advocated by several nongovernmental organizations to call a temporary halt to the sending of women migrant workers abroad as they were becoming the victims of various violations. The argument surrounding whether to continue sending women migrants abroad came to an end as the new minister for labor Alhilal Hamdi—apparently backed by the president—stated that the government will continue the policy. He argued that the crux of the matter is on stopping the migrants, but on the improvement of the preparation before sending them abroad. The minister also argued that the state could not afford to stop the migrants going abroad as the unemployment rate is alarming and the migrant remittances have contributed significantly to the national revenue.

Interestingly, Baharuddin Lopa, the current Indonesian ambassador to Saudi Arabia, publicly stated his agreement to stop sending domestic workers to Saudi Arabia as he saw more problems than benefits.[20] Thousands of women migrant workers in Saudi Arabia arriving through *Umroh* Visa and claiming to want to conduct pilgrimage illegally remained there and found jobs as domestic workers. The problem of illegal Indonesian migrant workers in Saudi Arabia is increasing as the numbers are snowballing. This problem has strained the relationship between the Indonesian and Saudi Arabian governments. An open clash occurred on November 23, 1999, between Indonesian migrant workers who were gathered in the Indonesian Consul Office in Jeddah with the Saudi Arabian police.[21] During the incident, which involved around 7000 Indonesian workers, around 50 Indonesian workers were arrested by the Saudi Arabian police. These Indonesians are the illegal workers waiting to be sent back to Indonesia as many of them do not have legal documents to work and are no longer able to stay in the country.

---

[19] Soeramsihono was assigned by former Labor Minister Abdul Latief as the director of the ministry's own social foundation (Yayasan Sumber Daya Manusia or Human Resources Foundation) among others to manage the overseas labor's insurance fee. The collected fee was allegedly misused to bribe the members of parliament to speed up the finalization of the National Labor Law in 1977.

[20] Lopa's statement will probably not affect the Indonesia–Saudi Arabia diplomatic relations, as he made the statement after his return to Indonesia where he will be sworn in as a Minister of Justice and Human Rights by President Abdurrahman Wahid (*Kompas*, March 17, 2001).

[21] As reported in *Republika*, November 26, 1999.

Migrant Workers in the Era of Reform

The issue of overseas migrant workers, although it received greater attention in the so-called *era reformasi* (the reformation era), is still generally confined in the margins of political development. The relatively marginal position of migrant worker issues also reflects the general situation of labor politics in Indonesia, which generally remains outside the mainstream politics. The efforts of the formerly suppressed labor union activists to enter mainstream politics are hampered by the fact that the political basis of the labor movement has been almost totally destroyed less than three decades of the New Order authoritarian regime.[22] The New Order propagation that the labor movement is always associated with the communist party has significantly depoliticized labor issues. In addition, the relatively narrow bases of industrialization processes in Indonesia have contributed to the small number of workers population that also limits their mobilization to play a significant role in national politics. Furthermore, the developmentalist ideology adopted by the economist–technocrats provides the New Order regime a quasi-scientific legitimacy for suppressing the rights of labor to establish their free organizations.[23] The political discourse on labor during the New Order regime was therefore encapsulated into the notion of human resources within the economic development paradigm advocated by the economist–technocrats. In this regard, the replacement of the word "*buruh*" to "*pekerja*" clearly reflects the gross attempt by the state to erase the possibility of developing class consciousness among laborers. The collapse of the New Order regime has released the state's restriction on labor forming free labor unions and entering national politics. Labor activists established several workers' political parties to articulate the political interest of labor; however, their political influence is still very frail.[24]

Since the early 1980s, compared with domestic labor issues, issues related to overseas migrant worker seemingly get more public attention, particularly as far as the press is concerned. Perhaps, the apolitical nature of the overseas migrant worker issues is less threatening for the government to suppress the reports. Only in 1985, when Sudomo became the labor minister, the migrants were prohibited from talking to the press regarding their problems as migrant workers in the Middle East. The press coverage on the plight of migrant workers, particularly those who work in the Middle East and Malaysia, has successfully placed their issues at the national level. In the case of migrant workers in the Middle East, the press specifically reported the

[22] On the elaborated labor politics during the New Order's period, see Hadiz (1997).

[23] In this regard, the New Order labor policy resembles the colonial policy that according to Breman (1990, 144) "from the start had proven more interested in cheap than in free labor."

[24] Among others, the new political parties established based on labor are *PBN* (Partai Buruh Nasional or Labor National Party) led by Muchtar Pakpahan, who is also the chairman of *SBSI*. Another minor party that is also pacing labor issues in its political format is *PRD* (Partai Rakyat Democratic or Peoples Democratic Party) under the chairmanship of Budiman Sudjatmiko. Both Pakpahan and Sudjatmiko were jailed in the latest period of the Suharto regime.

human rights abuses experienced by Indonesian female domestic workers. In the case of migrant labor to Malaysia, press coverage focused on the various fatal risk incidents as many migrants enter Malaysia illegally. The horrifying risk experienced by illegal migrants to Malaysia ranged from the sinking of their boats before landing on Malaysian's shore to the harrowing experience of enforced deportations by the authorities. It is partly because of the widely publicized press reports that the issues of overseas migrant workers have attracted significant public attention since the mid-1980s. The flourishing number of nongovernment organizations, which took up the cause of overseas migrant workers for their critical stand against the New Order regime, further placed the migrant worker issues at the center of the public political arena.

Under the systematic political subjugation of labor during the New Order regime, the articulation of labor interest through nongovernment organizations is therefore a very natural societal response to the existing political circumstances. Initially, the nongovernment organizations mainly assisted the migrant workers and their families in seeking compensation as many of them were unfairly treated by their labor recruiters, particularly concerning the violation of the contract agreement. The nongovernment organization's institutional assistance and legal assistance in the welfare issues have slowly established the trust and confidence of migrant workers and their families in the genuine comradeship provided by many nongovernment organizations. In further development, the nongovernment organizations have facilitated the scattered migrant workers in establishing their networks and organizations.[25] A significant number of local-level semi-formal migrant workers organizations were established in major areas where migrant workers mostly resided, such as in various districts in West Java, East Java, and West Nusa Tenggara. Many kinds of training related to the practical needs for improving their economic welfare and strengthening their political consciousness have also been provided by nongovernment organizations, particularly in voicing their needs and demands to the government recruiters' prospective employers. The sustained process of migrant worker empowerment has resulted in a significant increase in the migrant workers' political articulations, such as organizing protests to the state authorities on various migrant worker issues. The migrant workers also learned that the authorities would pay more attention to their demands if they could attract the mass media to cover their protests and complaints.

In a parallel development, the nongovernment organizations have also improved their assistance to the migrant workers by establishing a national umbrella organization to defend migrant workers. In 1997, *Kopbumi* (Konsorsium Pembela Buruh Migran, or Consortium in Defense of Migrant Workers) was established as an organizational forum for nongovernment organizations concerned with the improvement of the conditions of migrant workers. Along with the impetus that instigates the nongovernment organizations to establish their umbrella organization were the

[25] I would like to acknowledge the useful discussions on this issue with Yuniyanti Chuzaifah, an activist of *Solidaritas Perempuan* who is the Head of Komnas Perempuan.

polemics and debates surrounding the government proposal to promulgate the national law on labor (UU Pokok Ketenagakerjaan) in 1997. The content of the government proposal that was perceived as curtailing the various social, economic, and political rights of the workers stimulated wide-scale protests by labor and academics and nongovernment organizations. The government proposal also overlooked the specific problems and issues of migrant workers that triggered the migrant worker-related organization networks to start their campaign of protest against the proposal and further consolidated their effort to defend the political rights of the migrant workers. As protests were mounting, the government decided to pull out the new national labor law and stated that the draft should be revised. In such political circumstances, Kopbumi was established to immediately propose a special law for migrant workers lacking within the government's draft of national labor law. The existing policy on overseas migrant labor is generally based on various ministerial decrees that reflect the policy's ad hoc circumstances.

Another important development that also contributes to the increasing public attention on the issues of overseas migrant workers is the new discourse on globalization, which began around the mid-1990s. While the core of the discussion on globalization is still primarily dominated by the issues related to finance and trade regulations, it eventually also touched upon the issue of overseas migrant workers. In this context, the migrant workers who are largely unskilled, and many are undocumented, are generally passive participants in the margin of the playing field as the victims of the contest of globalization that is an arena only for "the big and the strong." International actors, primarily the transnational nongovernment organizations, such as ILO (International Labor Organization), IOM (International Organization for Migration), and other international NGOs—in many instances in cooperation with national governments and local NGOs—attempt to influence the law, regulation, protocol, and covenants concerning the international migrant workers in various ways. In Indonesia, the state's bureaucracy is generally very reluctant to adopt the international conventions that could improve the protection and strengthen the rights of the migrant workers. In addition to the fact that the issue of migrant workers is a low priority in the state's political agenda, the reason for this bureaucratic reluctance is probably because of the solid vested interests of the bureaucratic staff who benefited from the various illegal practices in the business of overseas migrant workers. The collusion between the bureaucracy and the labor recruiting agencies is an open secret and certainly hamper the genuine improvement of government regulations on migrant labor issues that are being encouraged by international agencies, such as ILO, to change the existing fraudulent policies on overseas migrant workers.

Other major actors playing a critical role in the industry of overseas migrant workers are traders and businessmen. Institutionally, these traders and businesspeople operate through various organizations, but most importantly the labor recruiters and suppliers' agencies (Perusahaan Jasa Tenaga Kerja Indonesia or PJTKI, member of APJATI). The crux of the matter is that the government is the sole authority able to provide the license for the labor recruiting agencies to operate. The lucrative business in overseas migrant labor has obviously constituted control on the association of labor recruiters and supplier agencies APJATI (*Asosiasi Perusahaan Jasa Tenaga Kerja Indonesia* or Indonesian Manpower Supply Association), although this has always been contested among the different business groups behind the agencies. The business groups dealing in labor exports, which interestingly have always been dominated by the labor suppliers to the Middle East countries, are largely owned by Arab ethnic groups.[26] The rivalries between different groups to control the association have occasionally resulted in conflicts during the election of their association chairman. The New Order's state maintained its political patronage by allowing only one association and giving its approval to the candidate who is most acceptable to government's interests, particularly the Minister of Labor. Recently, a new association—namely Association of Indonesian Migrant Workers Suppliers (Himsataki)—whose legality is strongly denied by APJATI, the existing association, was established. The reformation era apparently will not enable the post-Suharto's government to continue the policy for allowing only one migrant labor association.

Beyond the reach of the association's control, hundreds of labor exporting agencies have been operating their business in an unrestrained manner. While there are many legal labor recruiting agencies operating in Indonesia, there are also many that were operating illegally. In some cases, a combination of the two, in which legal labor recruiting agencies operate outside the legal procedures for recruiting the overseas migrant workers. The problem discussed here is related to unskilled labor because, obviously, skilled workers are more transparently regulated as the workers are mostly professionals and highly educated. Additionally, there are far greater numbers of unskilled workers, and majority of them are females with minimal education. In many cases, the recruitment process of these unskilled workers began in their home villages, which were often beyond the reach of the state's bureaucracy, except the local village authorities which authorized the statement of residency of the prospective overseas migrant workers. It is from these village levels that the business of overseas migrant workers started. The office of labor recruiter and suppliers' agencies located in nearby towns have relied on their intermediary agents who directly scout for prospective overseas migrant workers in the villages.[27]

The heavy demand for unskilled workers, such as domestic helpers for the Middle Eastern countries and plantation workers in Malaysia, coupled with the labor surplus in Indonesia, is a perfect match—they will continue to sustain the overseas migrant worker industry in Indonesia. Those who are politically powerless and have limited economic opportunities are vulnerable to exploitation by legal and illegal labor recruiting agencies, which firmly control the business industry of overseas migrant

[26] An attempt to interview these people on their business on labor exports to Saudi Arabia by the author in 1997 was not successful as they tend to avoid being studied. Their refusal to be interviewed reflects the degree of secrecy in their business activities, particularly in exporting female labor as domestic workers to Saudi Arabia.

[27] On the role of intermediaries in the business of labor migrants in East Java, see Spaan (1994).

workers. Furthermore, the continuing reluctance of the state's bureaucracy to improve the regulation policy on migrant workers will result in the burgeoning illegal movement without any legal protection. In these circumstances, the increasing political awareness among the migrant workers—as can be seen from the number of semi-organizations networks that are now developing with the assistance of nongovernment organizations—is crucial to strengthen the migrant workers' bargaining power *vis-à-vis* the state's authorities and the labor recruiting agencies.

The decision of the Abdurrahman Wahid government to implement the new national law on decentralization and regional autonomy (*UU Nomor 22 Tahun 1999 tentang Pemerintah Daerah dan UU Nomor 25 Tahun 1999 tentang Perimbangan Keuangan Pusat dan Daerah*) can prove to be a new opportunity to improve the policy on overseas migrant workers as the regional governments could be given the authority to regulate the movement of overseas migrant workers from their regions.[28] The fact that overseas migrant workers have significantly contributed to the regional government revenue will be a strong reason for the regional government bureaucracy to obtain control of their movements. Recently, a draft of the national law on the protection of Indonesian migrant workers and their families (UU Tentang Perlindungan Buruh Migran Indonesia dan Anggota Keluarganya) have been finally enacted by *Kopbumi* and proposed to the members of parliament and the government, as well as disseminated to those members of the public concerned about the fate of migrant workers. This is undoubtedly a landmark in the context of the existing dispute about influencing the regulating policy concerning the overseas migrant workers in Indonesia. The draft proposal also can be seen as a significant achievement by the migrant workers and their compatriots in what could be their long struggle in the politics for regulating the conditions of overseas migrant workers in Indonesia.

Concluding Remarks

The main argument of this chapter is that the state's response to migration is dominated by the political and economic forces of the ruling elites in the country. To better comprehend the Indonesian state's response to contemporary migration, it is necessary to understand how the ruling elite within the political and bureaucratic systems perceived the population issues, particularly the migration phenomena. In this regard, the legacy of past experiences of the state's response to migration, both

---

[28] Several provinces, such as Central Java, East Java, and West Nusa Tenggara, have begun to issue regulation policies concerning overseas migrant workers from the province. Among others, attempt to establish regional regulation on the business of overseas migrant workers in Central Java is the allocation of credit to the recruiting agencies in East Java. Another initiative has been shown recently by the opening of direct *Merpati* flight from Mataram, West Nusa Tenggara to Kuala Lumpur, Malaysia, in order to serve the heavy demand for transportation of migrant workers from West Nusa Tenggara who work in Malaysia.

during the colonial and postcolonial periods, has been crucial in shaping the perception, and, in turn, the responses of the current political and bureaucratic systems on migration and population issues in general. The state's response to international migration only began in the 1980s, after the state realized that both economic and political benefits could be sought through foreign employment policy. Reducing unemployment, which is perceived as one of the sources of domestic social and political unrest, and a way to acquire national revenue are the twin fundamental goals of the state's foreign employment policy.

The inward-looking character of the migration policy in Indonesia is reflected in the political–bureaucratic structure, which sets political stability and national integration as the national priority. Economic development planning, initiated by the economists–technocrats since the beginning of the New Order, apparently has not yet been integrated with foreign employment policy. Since the mid-1980s, international labor migration has become one of the national public issues due to heavy press coverage and stories about the desperate conditions of migrant workers abroad. Attention has generally focused on the lack of legal and institutional protections, primarily females engaged in generally unskilled occupations. The public also realized that many migrant workers moved overseas through illegal channels and without legal and institutional protections. In addition to the reported cases of mistreatment among the female migrants in the Middle East since the early 1990s, a series of fatal incidences experienced by illegal migrants who crossed the Strait of Malacca has been continuously reported by the press and mass media, bringing a lot of public attention to the fate of Indonesian migrant workers abroad.

The issue of overseas labor migration has become a topic of public concern. Several nongovernment organizations took up this case as their cause *vis-à-vis* the state that formally controls the policy and regulation. While the NGOs take a critical stance on behalf of the powerless migrant workers, other nonstate actors, namely the labor recruiters and suppliers, have long been very influential in the business of overseas labor. These businesspeople have strongly controlled the movement of workers abroad, both as a mediator with the state or by unlawfully assisting the prospective migrants. The tension between the actors involved in the attempts to influence the regulations on overseas migrant workers is slightly changing after the collapse of Suharto's regime as the state is no longer able to continue suppressing the strong demand to provide a legal basis for the protection of the overseas migrant workers. Nevertheless, the more fundamental problems currently facing the state and the nation to consolidate the disillusioned democracy and recover the national economy would probably hamper any chances of instituting a clear policy on overseas migrant workers soon. With the state being incapable of delivering institutional and legal protection for overseas migrant workers and the likelihood of increasing demand for unskilled overseas migrant laborers, the result will be the continuance of burgeoning problems surrounding illegal human trafficking with all its social repercussions. As such, a concerted effort is urgently needed in which the state and nonstate actors, both at the national and international levels, tackle the problems seriously.

Chapter 9
Internal–International Migration Divide?

Introduction

People are constantly on the move, transgressing social and state boundaries, seeking resources for their betterment of life. Indonesian people are no exception; they move either voluntarily or under the state's transmigration program, leaving behind their village or hometown to cities, other islands, and even abroad into the neighboring countries and faraway lands. Migration reflects human initiatives and responses to the changing nature of society and economy in their own country and beyond. This paper attempts to show that the conventional divide between internal migration and international migration in Indonesia is not always the case as has been demonstrated by data and information from some studies on migrant workers.[1] It argues that internal and international migration is a continuum and should be seen as interconnected phenomena.

This chapter is divided into four parts. The first part is an overview of migration and its contribution to the formation of Indonesian society. The second part will look specifically at migration as a response to differences in spatial development. The increasingly important migration phenomena of cross-border migration will be elaborated in Part III. In the fourth part, the interconnection between internal migration and international migration will be drawn.

Draft paper for a conference on "Crossing Borders, Traversing Boundaries: Bridging the Gap between International and Internal Migration Research and Theory," October 13–14, 2011, Asia Research Institute, National University of Singapore.

[1]The 2010 population census contains information on the return international migration, and those are the people that resided overseas 5 years before the census. The nationwide data on these former international migrants provide an opportunity to trace back their migration history including their movement within the country before their movement abroad. The intention to use census data to support the argument of this paper, unfortunately, cannot be done as the processing on migration data still underway and not available for analysis.

R. Tirtosudarmo, *From Colonization to Nation-State*,
https://doi.org/10.1007/978-981-16-6437-3_9

Migration and the Formation of a Society

Indonesia, an archipelago located between Asia and Australia, is currently the fourth largest country globally in terms of population size. According to the latest census (2010), the population is 237 million, geographically distributed into 34 provinces and more than 450 districts. It is estimated that more than six million Indonesians live abroad, mostly female, as domestic migrant workers. Ethnically and religiously, Indonesia is a diverse country, reflecting a long process of social transformation that continuously took place through human movement in this archipelago. The Dutch historians, Vollenhoven, and van Leur as quoted by Hugo (1980, 97) summarized patterns of migration during precolonial Indonesia as being of three main types: (1) colonization by large groups of migrants from one region who settled in another region, (2) migration of individuals, mainly traders, who settled in port cities, and (3) establishment of authority in foreign regions. From the sixteenth century up until the latter part of the nineteenth century, there was a gradual increase in the number of Europeans, whose activities had much influence on the pattern of migration in Indonesia.

According to the 1930 population census, 11.5% of the total indigenous population of Indonesia was living outside their district to birth. It was interesting that many Java-born Indonesians were enumerated in the other islands. Pelzer (1945a, b, 189) calculated that 967,000 Javanese, 120,000 Sundanese, 16,000 Madurese, and 39,000 other Indonesians of Java ancestries were found in the other islands, and not all of them were Java-born. Before independence, two factors contributed to the movement of people from Java to the other islands. First, the increasing Dutch plantations in the other islands where labor was scarce led planters to recruit people from Java. The first group recruited was Chinese coolies, but later Javanese and Sundanese workers were also recruited. Secondly, in 1905, the Dutch government considered Java to be overpopulated and introduced its colonization program to resettle people from Java in the "Outer Islands." However, the concentration of colonial activity in Java also led to several growing urban centers, such as Batavia, Surabaya, and Semarang. According to Pelzer (1945a, b, 175), these cities attracted people from the crowded interior due to the employment opportunities they offered. Hugo (1980, 114) argued that migration played a key role in the growth of urban centers during the colonial period. Although a significant proportion of urban dwellers were immigrants from overseas, in-migrants from rural areas and other islands made up a majority among the indigenous urban population. During the colonial period, besides movement due to the impact of colonization, there was also large-scale migration by people from the highly mobile ethnic groups, such as the Minangkabau, Bugis, Banjarese, and Makasarese.

During World War II (1942–1945), Indonesia was occupied by the Japanese. Little is known about population movement because no data were available during this time. The only evidence was that the Japanese initiated a forced labor *(romusha)* recruitment system, which resulted in the sending of many Javanese to the other islands or other countries in Southeast Asia; many of them never returned. After independence, the Indonesian government reinstituted the colonization program under *transmigration,* with their first transmigrants settling in Lampung, South Sumatra, in 1950. From the early 1950s, anthropologists and sociologists paid more attention to the migration of selected ethnic groups. The principal theme that emerged from these studies about the mobility of ethnic groups was that the mobility patterns were due to a tradition of migration or traditional pressures within these ethnic groups. These studies tend to consider that the forces behind the migration of these ethnic groups are more or less constant. Forbes (1981, 61), however, argued that forces producing this ethnic migration were constantly changing. Some forces (e.g., political unrest) were only significant at certain periods. Some others, like ecological, demographic, economic, educational, and urban attractions, increased steadily in significance, whereas the remaining forces (geographical and social–systemic) remained constant throughout the period or became insignificant.

McNicoll (1968, 37), based on the 1961 census birthplace data and the 1964–1965 National Sample Survey, found that South Sulawesi showed the greatest internal mobility, followed by North Sumatra and South Sumatra. The last two regions were also prime recipients of migrants. Regions of particular stability were West Nusatenggara and Bali, while Central Sumatra was an important out-migration region. Due to urban migration to Jakarta, the proportions of out-migrants from West and Central Java were comparatively large. In 1961, rural Sumatra, Kalimantan, and Maluku contained the largest "external" population. In Sumatra and Kalimantan, the in-migrants were transmigrants mainly from Java, while in Maluku the majority was from Sulawesi. At the other extreme, the rural areas showing the fewest in-migrants relative to their population and the least internal mobility were in East and Central Java, Nusatenggara, and Central Sulawesi. Besides seeking a better life in urban areas, rural to urban population movement is caused by insecurity due to political instability in some rural areas, most notably the Darul Islam rebellion in West Java.

According to the 1971 census, nearly 5% of the total population had migrated from their province of birth, while 1.2% had changed their province of residence more than once. The most important stream of migration (nearly 40% of the total net lifetime migration) was that flowing into Jakarta from all other provinces. About four-fifths of this migration was from other provinces in Java. Although the population of Jakarta was one-third of the urban population of Java, around two-thirds of all migrants to urban areas ended up in Jakarta (Sundrum 1976, 90). According to Speare (1975, 79), the second largest stream of migration was that flowing into Lampung, Jambi, and Riau. The percentage born outside those provinces was 36.16 and 13, respectively. The rest of the interprovincial migration was to provinces experiencing high rates of economic development in recent times, such as East Kalimantan, where there were timber and oil industries.

The most comprehensive study about voluntary migration of highly mobile ethnic groups is probably the work of Naim (1979). He made an intensive study about the Minangkabau migration (*merantau*), exploring the why, where, and its consequences upon the place of origin and the place destination. According to Naim, *Merantau* is a unique type of migration behavior. It is a voluntary movement, usually leaving one's cultural territory—whether for a short or long time—aimed at earning a living or seeking further knowledge and experience. In the migration literature, *merantau* is considered temporary or a circulation as the migrants intend to return home. Additionally, Naim (1979) compared the migration volume of some of the most highly mobile ethnic groups since 1930. In 1930, the percentage of migrants among the Bawean was the largest (35.9%), followed by the Bataks (15.3%), the Banjarese (14.2%), and the Minangkabau (11%). The projected migrant population in 1961 showed that the Bawean is still the most mobile (31.6%), followed by the Minangkabau (31.6%), the Bataks (19.5%), and the Banjarese (12.2%). In 1971, the projected Minangkabau migrants were 44%, probably the highest among the ethnic groups in Indonesia.

The island of Sulawesi is also the homeland of several of Indonesia's most peripatetic ethnic groups. The Bugis, the Makasarese, and the Torajan are the three largest ethnic groups in South Sulawesi. Lineton (1975a, b) showed that out-migration has a long history among these ethnic groups. The fact that the Bugis have been migrating for centuries predisposed other Bugis to leave their native land, noting that the Bugis colonies abroad are sources of information about more favorable economic opportunities elsewhere and often provide both financial assistance and moral assistance to the prospective migrant. The success of many former migrants arouses a spirit of emulation in those who have remained behind. Migration is seen to be a means—the only means for many people—to achieve wealth and a higher social status (Lineton 1975a, b, 190–191). There is also substantial Bugis movement within South Sulawesi, including seasonal circular migration between villages and the provincial capital city of Ujung Pandang. Much of this rural–urban movement also has a seasonal rhythm and involves the Makasarese and the Torajans.

For three decades (1971–2000), the population of Indonesia has rapidly grown from 119.3 million to 203.9 million. Concentrated in one island, Java now comprises around 60% of Indonesia's total population, recording a decrease from 73.8% in 1971 to 60.8% in 2000. Apart from Jakarta and West Java, other provinces in Java have relatively low population growth, presumably indicating a high rate of out-migration, especially from Central and East Java provinces. The movement of people from Java to the other islands, most notably through emigration policy under the Dutch, is continued by the Indonesian government under transmigration policy. Apart emigration and transmigration policies, the Javanese, being the largest ethnic group in Indonesia, is inherently the most mobile ethnic group. In 2000, the Javanese demographically dominated Java and in provinces outside Java, such as Lampung, North Sumatra, East Kalimantan, Jambi, and South Sumatra. Indonesia's national motto, unity in diversity, reflects the process of social transformation in the making of the Indonesian nation.

Migration and Regional Disparity

Many improvements have been made in almost every aspect of social life during the Suharto era. In the first five-year development plan (1969–1973), the government emphasized the development of infrastructure and communication, i.e., roads, railroads, harbors, and airfields. In the social sectors, the growth in education was also dramatic. While only about half the children at primary school age were enrolled in schools 15 years ago, now many more children at that age have access to schooling. Each village with a population of around 2000–3000 people has at least one primary school, and many have more than two. Health facilities have entered the 50,000 villages with the provision of integrated health services for every subdistrict. At the beginning of the second half of the 1970s, satellite communication was introduced. Later, the government distributed television sets in every subdistrict. The combination of the increase in education, health services, and the availability of transportation and communication facilities has raised the hopes and aspirations of rural young people, which resulted in the movement to urban centers.

Data on recent migration clearly show that the number of migrants for 1985–1990 is larger than the previous two periods (1975–1980 and 1980–1985). The provinces with a higher rate of recent migrants increase during the 1985–1990 period are Aceh, Riau, Bengkulu, West Java, Central Java, East Java, East Timor, Central and East Kalimantan, South Sulawesi, Maluku, and Papua. In contrast to other parts of the country, the increasing number of recent in-migration to provinces in the eastern part of Indonesia was already occurring in the 1980–1985 period. As the government capacity to sponsor population movement through transmigration program became limited due to the collapse of oil prices in mid-1980s, most recent migrants in the 1985–1990 periods were most likely to be voluntary migrants. The various economic reforms implemented since the mid-1980s to boost foreign investment have resulted in the opening of job opportunities in many places, such as Batam Island in Riau, East Kalimantan, and Papua. A significant indication that most recent in-migrations consist of spontaneous or voluntary migrants is their destinations, which are mostly urban areas compared to rural areas. The process of labor reallocations from low-income agriculture to higher productivity nonagricultural employment accompanied the population mobility change from assisted to voluntary migration.

Internal migration is a critical aspect of Indonesia's provincial economic growth, not only because of its magnitude but also because it has been the government's policy in encouraging economic convergence in Indonesia. Vidyattama (2008), based on the 2000 census data, calculated that around 10% of the Indonesian population was not born in the province where they reside, and around 3% resided in different provinces again 5 years earlier. The migration flow from low-income provinces to medium- and high-income provinces was still significant in Indonesia during 1975–1980 and 1985–1990, comprising approximately 26% of all migration flows. The lower number of migrations from lower-income provinces was mainly due to the low emigration rate from East and West Nusa Tenggara, despite the low population in low-income provinces being another reason as well. Human capital must play a role in determining the low migration rate from West and East Nusa Tenggara, since the migration rate from Maluku or Southeast Sulawesi (where human capital is higher) was still much higher than for the two Nusa Tenggara provinces. However, the number dropped to only 9.3% during 1995–2000. This drop resulted from the large increase in the migration among medium-income provinces, partly because of the 1997–2000 crises and conflict. As expected, the opposite direction of this flow (i.e., from high- and medium-income provinces to low-income provinces) was low at below 15% of all migration, except during 1975–1980, when the flow from medium- to low-income provinces alone was 14.9%. The transmigration program was the main reason for this high flow to low-income provinces in that period.

Human capital is another critical factor in determining migration's impact on growth. As discussed, the human capital in Indonesia of both out-migrants and in-migrants is always higher than the average workforce human capital. As a result, migration will have a clear positive impact on human capital if the human capital level of immigrants is higher than in-migrants and the province has positive net migration. In contrast, the impact will be negative if the province has negative net migration, and in addition, the human capital of out-migrants is higher than in-migrants. Besides these two conditions, it is hard to measure whether migration will increase or decrease a province's human capital level. Only a few provinces have a clear positive or negative impact from migration on their human capital. West Java was the only province with higher educated in-migrants than out-migrants for the entire 1980–2000 period, and thus, migration was highly likely to boost human capital. It also had positive net migration during 1980–2000. Jakarta and East Kalimantan were provinces with higher educated in-migrants than out-migrants until 1995, but Jakarta also had negative net migrants after 1985 that may have caused human capital depletion.

Riau, Bali, and Yogyakarta all had positive net migrants and higher educated immigrants than emigrants in most periods, especially after 1990. Yogyakarta plays a special role as one of the centers for education in Indonesia besides Jakarta. As a result, Yogyakarta has always had the highest average education for both out-migrants and in-migrants. However, in contrast to other periods, the level of education of out-migrants was higher than for in-migrants during 1985–1990. That is most likely related to the 1986–1992 industrialization period in Indonesia that attracted university graduates to other Java provinces, especially those from Yogyakarta. Riau and Bali had developed manufacturing and tourism industries, respectively, from the end of the 1970s. However, during 1980–1985 and 1985–1990, out-migrants level of education from these two provinces was still higher than for in-migrants.

The trend and pattern of human movement in the eighties continue in the 1990s and 2000s. With the substantial reduction in the government capacity to move people under transmigration policy, migration significantly develops a function of labor market economy. In situations where the government can only play a minimum role in directly influencing human mobility, labor recruitment agencies become more influential alongside personal networks. Labor recruitment agencies, or other kinds of mediators and brokers, emerged as the major players in the lucrative migration industries of overseas contract workers. The surplus of labor in densely populated areas, particularly Java, Bali, and Lombok, makes them to become a source of labor for other places within Indonesia and beyond. Internally, the opening economic opportunities in some places like Papua, and the fact that the local population has relatively low human resource endowment, constitute the structural inequalities that create push and pull factors for people's movement.

Speaking of the human movement, Papua should particularly be given serious attention in the future. Many researchers have reported the increasing communal tensions in the region as the locals perceived migrants as threatening their resourced economy and cultural identity. In-migration to Papua, while in many ways reflects the work of labor market forces, also represents local demands for more political recognition—that is precisely what pushes locals to perceive migrants as threats. A comprehensive political economy approach, in which freedom of movement shall not be jeopardized, is needed to resolve this current increasing communal tension in Papua.

Since the beginning, the economic development policies under the New Order government have been seen as a result of the increasing regional disparity. From a structural perspective, Titus (1978) and Forbes (1981) have observed the impact of unequal development on the pattern and trend movement of people between provinces. They argued that only a more egalitarian and decentralized policy that benefits both peripheral regions and marginal social groups could end the problem. In May 1998, Suharto's three decades' old New Order regime collapsed after failing to revive the country's economic fortunes. Following this, one of the new rapidly developed political landscape features was the devolution and decentralization of power and authority to the regional-level governments. In January 2001, a new regional autonomy law was formally implemented, full of promises to become, in Titus' words, a "more egalitarian and decentralized policy" benefiting both peripheral regions and marginal groups. As a matter of fact, inequality cannot be isolated within Indonesia's borders. People's movement indicated that economic inequality and uneven development have strong roots within the state's border and beyond the state's border.

While the movement of people from Java to the other islands continues to increase, big cities in Java, particularly Jakarta, Semarang, Bandung, and Surabaya, also continue attracting migrant from all over the place in Indonesia. The latest World Development Report by the World Bank argued the importance of the economics of geography. It emphasized the empirical evidence of agglomeration as the engine behind economic growth in the advanced industrializing countries in Western Europe, Northern America, and East Asia. In Indonesia's economic development, the expanding tendencies in the public infrastructure investments, like the current construction of the highways connecting big cities (Jakarta, Bandung, Semarang, and Surabaya)—mainly in Java, will accelerate the agglomeration process in Java. The human movement will be attracted by such a massive concentration of economic activities in Java. Rural to urban and interisland migration into Java will continue to be the major contributions to the agglomeration process in Java. While high economic growth might be achieved through agglomeration processes, as implicated by the World Bank's recent report, such development might also paradoxically reflect a continuation of social and economic inequalities between the rural and urban areas, as well as between Java and the other islands. The archipelagic form of Indonesian geography, and the plurality of its society, should perhaps be reconsidered by the state's planners. The archipelagic form of Indonesian geography, and the plurality of its society, should perhaps be reconsidered by the state's planners. That way, the agglomeration process in Java would not be counterproductive to the nation building and national integration processes.

Cross-border Migration as a Continuation of Internal Movement

International migration in the contemporary sense is a recent development in Indonesia. However, before its independence in 1945, a handful of people from Indonesia have had experiences in migrating to places that are now considered foreign countries, as far as Holland, Suriname, South Africa, Sri Lanka, Malaysia, New Caledonia, and Australia. The Javanese in the 1920s and 1930s were recruited by the Dutch as contract coolies to work in various export crop plantations in Suriname, New Caledonia, Vietnam, Johor (West Malaysia), and North Borneo (East Malaysia). The Ambonese from the Maluku Islands, for example, were employed by the Dutch army, followed them, and resettled in the Netherlands. Some Indonesians were exiled by the Dutch to Sri Lanka and South Africa, while few Indonesian nationalists were stranded in Australia after the Allies freed them from Digul's political exiles camp in the southeast of West Papua. In a more voluntary ethnic mobility, a wave of Minangkabaus migrated to Negeri Sembilan in the Malay Peninsula and has now become Malaysian.

In the Southeast Asian contexts, the Malay Peninsula, Singapore, Sarawak, Sabah, Southern Mindanao, and Indonesia are known as "the land below the wind" or the Malay World. People's movements in the Malay World are common and continuous, although the decolonization state boundaries are erected and thus divided people into different nationalities: Indonesian, Malaysian, Singaporean, Sarawakian, Sabahan, and Filipino. However, cross-border movements are often seen as if it was a new phenomenon where the border is older than the movement of people (Ludden 2003). Again, in this context, international migration is a label for migration behavior that the people traditionally practice without any awareness of the existing state boundaries. The current disputes between the Philippines and the Sabah state government over many Filipino migrants in Sabah are just a case in point. The claim that Sabah was part of the Sulu's sultanate provides a historical legitimacy for the people from the southern part of the Philippines to stay in Sabah. The prolonged and unresolved problem of Filipino migrants in Sabah now resulted in the ambiguous legal status of the second and third generations of the Filipino refugees in Sabah (Kassim 2009). The people from the island of Sangir and Talaud in North Sulawesi also experienced similar ambiguous situations. They now cultivated land in the southern part of Mindanao and are considered as "undocumented citizens" by the Mindanao government (Ulaen 2011). Another issue concerning migration in border areas relates to the maritime ethnic minorities, such as the Orang Suku Laut who wandered over Riau Islands, Malaysia and Singapore, and the Bajaus, whose territories comprise Sulawesi, Sabah, and the Southern Philippines. These sea-based mobile ethnic groups constantly negotiated with different state apparatuses to survive and secure their livelihood (Chou 2006). These examples of people's lives in the border areas reflect the unsettled divide between internal migration and international migration in this region. International migration is an official imposition to a social reality of a human continuum movement. In the current migration, such a continuum is then ruptured in the hand of the migration analysts.

In the last two decades, the rift division in the literatures between internal migration and international migration is increasingly getting wider—such a gap results from the fact that knowledge on migration mainly comes from the north-rich industrialized countries. The influx of migrants from the south-poor countries has created problems in the north-rich countries, from the UK and Europe to Australia and New Zealand, Japan, Canada, and the USA. The influx of immigrants urged the governments in the north to devise immigration control to respond to such increasing problems. Government-financed think tanks and university research centers were then asked to conduct studies on international migration, particularly on immigration's social, economic, and political implications. The perception of immigration as a threat cannot be exaggerated as the underlying concern that attracted many researches on international migration.[2] The growing fear of the "Muslim invasion," especially in Europe, strengthens the nationalist–populist movements and political parties that were openly against immigration (Schierup-Carl et al. 2006).

In Indonesia, media reports on the plight of migrant workers, especially the exploitation and abuse experienced by female domestic workers, placed and increased the importance of international migration in public discussions. The exploitation and abuse of domestic workers in the receiving countries, most notably Saudi Arabia, Malaysia, and to a lesser extent Singapore, captured by mass media and sparked controversies that pushed the government to properly manage the policy on overseas migrant workers. Again, there is certainly a bias toward female domestic workers; it overlooked the fact that Indonesian migrant workers also suffered from abuse and exploitation. The public perception that perceives Indonesian workers abroad as good might be accurate for professionals and more formal occupations,

[2]The title of a book edited by Myron Weiner (1995a, b), a long observer on the politics of migration, namely "Global Migration Crisis," epitomized such a deep concerned of the crisis implicated by immigration in the north. A more balanced view, yet mostly about international migration, perhaps is represented by Castles and Miller (1993) on their book entitled "Age of Migration."

such as engineers, nurses, and sailors. However, it is certainly not the case for construction and plantation workers, particularly in Malaysia. The remoteness of location perhaps contributes to the lack of attention from the media to expose the exploitation and abuse of plantation workers in Malaysia, both in West Malaysia and in East Malaysia.

In terms of policy, the turning point in overseas migration occurred in the early 1980s when the government realized the need to regulate the increasing flow of overseas labor, most notably women—to Malaysia and the Middle East—as domestic workers. Private agencies partly assisted the flow of migrant workers abroad at the time, while the rest was moving voluntarily as part of the social and family networks. As stated earlier, the trend and pattern of internal migration in the 1990s and 2000s constitute a function of the labor market economy as the state's capacity in sponsoring migration is declining since the mid-1980s. A paradigm shift in migration occurred from official to unofficial type of human movement. In situations where the government only plays a minor role in directly influencing human mobility, labor recruitment agencies are becoming more influential alongside personal networks. Labor recruitment agencies, or other kinds of mediators and brokers, emerged as the major players in the lucrative migration industries of overseas contract workers. In this situation, the surplus of labor in densely populated areas, particularly in Java, the southern part of Sumatra and Sulawesi, and Lombok, logically becomes the source of labor for other places within Indonesia beyond.

Labor surplus is only one factor in overseas migration; other factors, most importantly labor recruitment agencies and family networks, are also very influential. The flow of the Florenese from Eastern Flores from the early 1950s to Sabah is an example of how family and kinship networks nurtured the Florenese migration (Tirtosudarmo 2006). Flores, Timor, and Sumbawa islands in Eastern Indonesia became the source of overseas labor. That is because their relatively poor economic conditions pushed productive age populations to migrate, and the increasingly active role of brokers and mediators made them recruit and channel more labor into working overseas (Linquist 2010). Internal and international migration occurred side by side as both individual and institutional responses to the job market and other economic opportunities. The state's involvement in the labor movement abroad accelerated the flow of migrant workers and created new problems as the state could not formulate a proper policy to manage the flows of overseas migrant workers. The high demand for domestic workers increased steadily, creating what is popularly known then in the migration literature as a migration industry.

The media attention to the plight of migrant workers and the stronger voices of NGOs advocating human rights protection contributes significantly to the dominant public discourse on international migration. The development of media attention and the NGOs' movement on migrant workers cannot be isolated from the increasing concern at the global level. Migrant worker issues captured the dilemmas and paradox of globalization immediately. The popular slogan of "borderless world" epitomized globalization is merely a mocking for migrant workers from the poor countries and hypocrisy of the rich countries. While information, capital, and trade are moving freely from the north to the south, migrant worker movement from the south to the north is strictly controlled. The securitization of migration following the 9/11 event further tightened the immigration policies in the north-rich countries.

Undoubtedly, international migration and its related issues have dominated the literature leaving behind the internal migration statistically much higher in proportion than international migration. Attempts by scholars to bridge the gap between internal migration and international migration are interestingly very few. The most recent and comprehensive attempt to integrate the approach to internal and international migration is perhaps by King and Skeldon (2010). They argued that the divide between internal and international migration emanates more from the different "traditions" of these two types of migration. By "tradition," they mean the approach, the methodology, and the policy agendas surrounding the two types of migration. They also pioneered the attempt to bridge the gap between the two "traditions" by proposing a model based on Mexico's migration experience. The Indonesian migration experience, as indicated by several research works, in many instances supportive of the propositions put forward by King and Skeldon (2010) on the internal–international migration linkages.

Three recently submitted Ph.D. theses on overseas migrant workers trace the surrounding livelihood of the return migrants in their place of origin. Although these three studies are not explicitly looking at the connection between internal migration and international migration, their qualitative data show that internal migration precipitated the international movement of labor in their study areas. Primawati's (2008) study in Purwodadi District, Purworejo Regency in Central Java, found that "the productive age groups commonly migrated to other places, they habitually move, permanently or circularly, to other villages, and moving from one city to other cities, and even to overseas." Rosadi's (2010) study in Juntinyuat District, Indramayu District, in West Java, confirmed that migration had become an important feature of the people. Rosadi noted that "the number of migrant workers, male and female, originating from Indramayu Regency to other cities or abroad steadily increased." Setiadi's (2010) study in Temon District, in Yogyakarta Special Region, argued that "social, economic and cultural life of the community has been shaped by the dynamic of a long history of migration experience by the people in the Temon District." These three studies concurred the importance of asphalt road infrastructures in connecting their district studies to the capital cities, in this case Purwodadi and Temon districts to Yogyakarta and Juntinyuat districts to Jakarta. The free access for villagers, both males and females, to move from their villages to the capital cities provides structural support for migration as an essential habit for the people in these two districts, both internally and internationally. Another indication of the potential connection between internal migration and international migration is shown in the set of data on internal migration collected in a survey in 2011 by a collaborative team from ANU and Gadjah Mada University in four cities: Tangerang (West Java), Medan (North Sumatra), Makassar (South Sulawesi), and Samarinda

(East Kalimantan).[3] This survey was aimed to understand the latest development of internal migration by interviewing migrants in the four selected cities that geographically represented the western, middle, and eastern parts of Indonesia. It was found that around 40 respondents in the survey, apart from their internal migration experiences, have also experienced international migration.

The tradition of migration and the regional disparity, internal and internationally, constantly contribute to people's movements. The state and the labor recruitment agencies are tapping the benefit from the people on the move by providing access that is open by the labor market economy in Indonesia and beyond.

Conclusion: The Internal–International Migration Nexus

Migration has become one of the main features for social transformation in Indonesia. As part of insular Southeast Asia, Indonesia is the homeland of several ethnic groups known as the seafarers, such as the Bugis, Makasarese, Butonese, Mandarse, Banjarese, and Madurese. With the improvement of land transportations, practically no more isolated communities as people could move between places for various reasons in which economics always plays an important role. The disparity in economic development between rural and urban areas, and between Java and other islands, constitutes the underlying structural factors that eventually push and pull people to migrate, whether permanently or temporarily.

The division between internal migration and international migration from the Indonesian experiences was partly a result of the creation of the nation-states following the process of decolonization in the mid-1940s. The creation of state boundaries immediately divided the movement of people into internal and international migration. The movement of people common in the border areas in Riau Islands, in Kalimantan, and islands of Nunukan and Sebatik, Sangir, the Talaud Islands in North Sulawesi, in Papua, and Timor suddenly became cross-border and fell into the international migration category. The rising issues related to citizenships, irregular migration, and human trafficking are among the current social and political implications of the imposing state's legalistic categorization on the human flows in the border areas.

The movement of workers, primarily women, to other countries cannot be seen as a separate phenomenon from the experience of migration within the country. Working as domestic helpers for rural women is common in middle and upper-class urban households; it provides a sociocultural background that legitimizes the increasing number of women working abroad as domestic workers. Migrations are becoming a habit for the people as several studies in the sending areas of domestic workers have indicated. In this regard, state borders are eventually becoming irrelevant for these

[3] I would like to thank Chris Manning (ANU) and Mujiani (Gadjah Mada University) for allowing me to get access to their migration survey data.

women as the imagination of households, hypothetically, could be essentially similar—no matter whether here or there, in Indonesia or overseas. Labor recruitment agencies or other kinds of mediators and brokers emerged as the major players in the lucrative migration industries of overseas contract workers.

The Indonesian migration experience supports other counties' experience that geographical distance is no longer relevant in defining internal or international migration. As Indonesia is concerned, the current hegemonic discourse that divides internal and international migration is a fallacy. The division results from a bias emanating from an imperfect understanding of migration processes influenced by the media exposure and the global research that tends represents the interest of rich industrialized countries—immigration is the problem of economically rich, not the poor countries. The domination of knowledge production emanating from the immigration policy agenda in the north industrialized countries produces an imperfect migration processes. A balanced perspective on internal–international migration as a social and cultural continuum in sending countries is critical for policymakers to formulate a coherent migration policy.

Part IV
Human Flows, Identity Politics, and Nation-State

THIS PART ATTEMPTS to understand the linkages between population mobility and conflict, specifically in conjunction with the increasing sentiment of "primordialism" in Indonesia, from the perspective of political demography. The structural contexts, a complex mixture of demography, geography, economy, and culture, significantly induce the pattern and trend of the recent and future dynamics of the political changes and social transformation. In this regard, an understanding of the nexus of demography and politics is vital to shedding light on the social and political ramifications of the geographic population mobility in Indonesia. From a historical perspective, the current situation marks a turning point in the population mobility trend, which is characterized by the phenomena of a strong return migration stream to Java as well as to other places of origin. The crucial role of ethnic mobility in the resurgence of politics of identity no doubt has created a climate of uncertainty in the current political reforms and the realization of democracy and civil society in Indonesia.

Firstly, in Chap. 10, the role of Bugis migration into one of their major destination areas, East Kalimantan, will be appraised. As one of the major migrating ethnic groups in Eastern Indonesia, the Bugis have significantly influenced the social, economic, and political development in East Kalimantan. Secondly, Chap. 11 will analyze the relation between migrants and the local population in Lampung province in the southern Sumatra. As shown in the first part of the book, particularly Chap. 1 Lampung is the first destination of the colonization program of the Dutch in the beginning of the twentieth century. Most of the migrants in Lampung were the Javanese that now become the majority. As the second, third, and fourth generation of the Javanese born in Lampung, a question arises whether they could be called as the stranger anymore as in fact they are also becoming the new locals. The last chapter (12) will discuss ethnic, geographic mobility and its consequences on the emergence of identity politics in Indonesia. This chapter illustrates the increasingly complex relationships between the notion of ethnicity and the nation-state's formation in Indonesia, following the implementation of the new decentralization laws in 2001. The chapter ends with the call for reinventing civic nationalism, which

Chapter 10
Ethnic Mobility: The Bugis in Samarinda, East Kalimantan

Prior and following the step-down of Suharto in 1998, Indonesia has been rocked by a series of violent conflicts. The first large violent conflict broke out in December 1996 in the Sanggau Ledo subdistrict in Sambas, West Kalimantan, between Madurese migrants and local Dayaks. Following the forced resignation of Suharto in May 1998, social, religious, and ethnic conflict has become a seemingly permanent feature of the post-Suharto political landscape. In the series of communal conflicts, migrant populations have always been involved. In many cases, entire migrant communities have been transformed into victims forced to leave their settlement areas, such as the Madurese in West and Central Kalimantan and the Butonese in Ambon, Maluku (Spice Islands). The involvement of migrants in conflict with the local population of differing ethnic backgrounds exposes the serious question of the relationship between migration and ethnic conflict. A recent study on the anatomy of violence in Indonesia in the last 10 years found that communal and separatist violence caused the greatest number of fatalities, accounting for 77% and 22%, respectively. While social violence was not uncommon during the Suharto regime, social violence and fatalities increased dramatically during the period of transition to democracy, reaching its peak in 1999–2000. The study also shows that among communal violence, ethnic religion migration-related violence alone contributed to around 52% of total deaths in social violence.[1]

This chapter is about the migration of the Bugis to Samarinda, the capital city of East Kalimantan Province. The Bugis are the dominant ethnic group in South Sulawesi and, alongside the Minangkabau of West Sumatra, the Madurese from Madura Island, and the Banjarese of South Kalimantan, one of Indonesia's major ethnic groups with a strong tradition of migration. The Bugis migrated with their ships and settled almost all areas of the archipelago, in several places constituting a

[1] See *Anatomy of Social Violence in the Context of Transition: The case of Indonesia 1990–2001*, by Mohammad Zulfan Tadjoeddin (Jakarta: United Nations Support Facility for Indonesian Recovery UNSFIR), March 2002.

R. Tirtosudarmo, *From Colonization to Nation-State*,
https://doi.org/10.1007/978-981-16-6437-3_10

dominant ethnic group. East Kalimantan is one of the prime destinations for Bugis settlement. Beside the Bugis, East Kalimantan is also a destination for Banjarese from South Kalimantan, Torajans from Central Sulawesi, and Florenese from East Nusa Tenggara. The state-sponsored transmigration programs have also targeted East Kalimantan as a major destination since the 1970s.[2]

The relatively peaceful situation in East Kalimantan, in comparison with the explosive conditions found in West and Central Kalimantan, shows that the migrant and local populations do not invariably conflict, although social tensions are always a natural part of the adjustment process between people from different ethnic backgrounds in East Kalimantan. The experience of the Bugis migrant community in Samarinda and East Kalimantan, overall, demonstrates that they have not only been successful in adjusting to the new place but also have contributed significantly to establishing a new multiethnic society strongly dominated by Bugis culture. The Bugis settlement in East Kalimantan provides a strikingly different case from the experience of the Madurese in West and Central Kalimantan, who have been constantly in conflict with the local population. The domination of the Bugis, in the fields of both economy and politics, has insulated East Kalimantan from these difficulties simply because the Bugis colony is acknowledged as dominant and well-established.

Among demographers in Indonesia, ethnicity is something remote, especially when we talk about migration in Indonesia. Demographers tend to ignore the ethnic and cultural dimensions of the populations under study. It seems that ethnicity is a subject to be left to anthropologists. The ethnic blindness of demographic study in Indonesia also results from the fact that aggregate data on ethnicity are just unavailable as questions on ethnicity have been omitted since the first census after independence. Only in the last census, in 2000, ethnicity was again part of the questionnaire.

A Brief Historical Background

According to Christian Pelras, a French anthropologist who has studied the Bugis for decades, Bugis settlements outside South Sulawesi have been established in the archipelago since the end of the seventeenth century. So, by the end of the eighteenth century, Bugis settlements can be found in Sumatra, such as in Bengkulu and in the Riau Islands, in the Malay Peninsula, in Kalimantan, among other places in and around Pontianak, Mempawah, Pulau Laut, Pagatan, Pasir, Kutai, Samarinda,

[2] Kutai District has become the main destination for transmigrants to East Kalimantan, receiving almost 50% of the provincial total. The official statistics from the Department of Transmigration show that the number of transmigrants resettled during the New Order period is as follows: Pelita I (1969–1973) 1775 families, Pelita II (1974–1978) 3311 families, Pelita III (1979–1983) 11,878 families, Pelita IV (1984–1988) 15,179 families, Pelita V (1989–1993) 16,525 families, and Pelita VI (1994–1997) 14,054 families (Tirtosudarmo 1998).

Bulungan, and Gunung Tabur. In the east of Makassar Straits, the Bugis settled in Palu Bay, in Donggala, Banawa, and Kaili, in Sumbawa, while in Java, the Bugis settled mostly in port cities, like Surabaya, Gresik, and Batavia. After the eighteenth century, the Bugis also settled in Indragiri and Jambi in Sumatra, along the west coast of Johor and Sabah in East Malaysia and around Lindu Lake and the southeast of Sulawesi. In the eighteenth and twentieth centuries, Bugis settlements were founded in Eastern Indonesia, including Maluku, Papua, and East Timor. Since the early 1980s, the three ethnic groups originally from South Sulawesi—namely the Bugis, the Butonese, and the Makassarese—have gained notoriety in Eastern Indonesia as the *BBM.* Tensions and conflicts with the local populations are quite common in Papua and East Timor since the early 1990s (Aditjondro 1986; Tirtosudarmo 1997).

The reason for Bugis migration, as for other ethnic groups, is always complex; there is no single explanation. On this, Pelras (1994, 319–320) said that the reason for Bugis migration is related to *"the resolution of personal conflict, an affront received, political insecurity or the desire to escape either unsatisfactory social conditions or undesirable repercussions of an act of violence perpetrated at home."* In this regard, Pelras has made an interesting observation that the reason for the Bugis migration is not always related to economic matters—usually a dominant perception in the literature on migration. Another characteristic that Pelras also mention regarding Bugis migration is the structural pressures that make each period apparently develop its own momentum in shaping the push and pull factors for the Bugis out-migration from their place of origin in South Sulawesi. Another Bugis expert, Lineton (1975a, b, 9), argued that the Bugis *"are peculiarly dependent upon their relationships with other regions for the maintenance of their society in its present form."* She further noted that *"As a result of political and economic pressures, the Bugis have, over a period of at least three centuries—become increasingly integrated within the wider socio-economic system of the Malay world."*

Lineton (1975a, b, 22) noted four characteristics of the Bugis migrant settlements. First, the destination is usually a less densely populated area. Second, their settlement is usually swampy lowlands that have a heavy rainy season particularly suitable for growing rice that will one day turn into a coconut plantation. At first sight, the Bugis settlement looks like an unattractive swamp area, but in fact, it represents a potentially very profitable ecological niche. The third characteristic of the Bugis settlement are its closeness to a seaport accessible with their boats, giving them a convenient place to bring their agriculture production to market. The fourth characteristic in choosing the settlement is the potential for future settlement expansion to adjacent areas.

A study of Bugis migration from Wajo shows that the largest out-migration flows occurred during the period of the Kahar Muzakkar rebellion between 1950 and 1965. The population of Wajo, particularly in the rural areas, faced a dilemma, as they had to stand between the rebels and the Indonesian military. According to the study in Kalola Village, located around 23 km north of the Wajo District seat Sengkang, the residents are caught between the rebels and the Indonesian military. This phenomenon shows the high level of insecurity and feelings of vulnerability experienced by the people of Kalola.[3] Out-migration has become their most rational alternative. The people of Kalola usually migrated to Jambi and Indragiri in Sumatra, Toli-Toli in Sulawesi, and Samarinda in East Kalimantan. To finance their migration, some sold their lands to their neighbors, who decided to stay in the village. They migrated in groups from the port of Makassar via Tanjung Priok harbor in Jakarta and then on to their place of destination in Sumatra. For the people who migrate to Samarinda and Toli-Toli, they boarded their boat in Pare-Pare. The massive movement has left Kalola almost empty. It was reported, for instance, that in Lawartanae—the main village of Kalola, out of around 300 households, only 30 households remain. Langkautu, which had 100 households, and Awatanae, which had 200 households, were found during the study to be deserted (Bisri, 1985, 16–17).

Social disruption and political chaos thus are strong push factors for massive Bugis out-migration from Kalola to the land beyond. South Sulawesi communities have long been characterized by insecurity. During colonial times, conflicts not only occurred between the Dutch and the "aru" or "karaeng" but also occurred between the "aru" and the "karaeng" themselves (Putra and Shri, 1989, 143–144). The end of hostilities following the killing of Kahar Muzakkar on February 31,965, apparently did not stop the flow of Bugis out-migration from Kalola, though the volume decreased some.[4] According to Bisri (1985, 19), it was difficult for people in Kalola to stay, as the neglected land became increasingly difficult to plow. Another factor that pushed the remaining people in Kalola to migrate came with the news that their fellow villagers had succeeded in the new settlements. Rice harvest failures in 1971 and 1972 caused by the drought convinced further numbers of Kalola to move to the new settlements.

Migration and Trade Networks

Bugis migration to East Kalimantan increased markedly after the 1970s, as shows the increased number of regular passenger's ship trips between Pare-Pare and Balikpapan and Samarinda. Among these ships are the large PELNI ships, such as the Tidar, Binaiya, and the smaller private ships, the Teratai, Tanjung Selamat, and Samarinda Ekspres. Other large PELNI ships were available from Makassar to Balikpapan: the Kambuna and the Kerinci. Bugis migration patterns, particularly as concerns the traders, have changed, as can be seen from patterns of capital

---

[3]See the study by Cik Hasan Bisri, *Kalola: Sebuah Desa yang Pernah Ditinggalkan banyak Penghuninya (Kalola: A Village Abandoned by Its People),* 1985.

[4]Lineton (1975a, b, 23) provides an interesting comment on Kahar Muzakkar's rebellion against the central government: "the most important aim of the rebellion has in fact been achieved: all the offices of government, from the lowest level of administration up to the provincial governor, are now filled by local born people."

ownerships. Migrants arriving in East Kalimantan around the 1970s generally obtain their capital from their own families or their close Bugis relatives. In the 1980s, the Bugis migrants in East Kalimantan, particularly Samarinda, known as "migran baru" (the new migrants), no longer rely on their families or close friends in acquiring the capital, but from any source whatsoever, not necessarily the Bugis.

The so-called old Bugis in Samarinda mostly resided in the place known as "Samarinda Seberang" a place located in the riverbank of Mahakam River in the city of Samarinda. These people are believed to be the descendants of the first generation of Bugis migrants in East Kalimantan. Some of these people are traders originally from Wajo District in South Kalimantan who still maintains close cultural and economic relations with Bugis elsewhere. A business trader, the author, met in a passenger ship from Makassar to Balikpapan explained that he dealt in numerous commodities. The commodities that he supplies depend on the seasons. Any commodity that is available either in South Sulawesi or in East Kalimantan may be traded, depending on the demand of the market on both sides. This phenomenon supports Pelras' observation that adopting the market economy is strongly characteristic of the Bugis migration strategy. According to this Wajo trader, if the market in Samarinda wants coconuts, he will send coconuts from Pare-Pare to Samarinda. If the people in Wajo are harvesting peanuts, he will go back to East Kalimantan with a full load of them. He admitted that in both Wajo and Samarinda, he already had a strong network consisting of fellow Bugis and Chinese traders.

For the Bugis traders, East Kalimantan is not just a destination within their trading network but a second home beyond their natural home, Wajo or South Sulawesi. In East Kalimantan, the Bugis traders also have long had a reputation as mediators and brokers for other traders in the border areas between Indonesia and East Malaysia. In Tarakan, a town close to the border, for example, the Bugis are well known as traders who bring various commodities, including electronic appliances and cloth, from the Tanah Abang market in Jakarta. From a cursory observation made by the author in Pare-Pare harbor, ships heading to Tarakan and Nunukan (the closest town to Sabah, East Malaysia) are loaded with rice coming from various places in the South Sulawesi, particularly from the well-known rice-producing district of Wajo. It is said that rice from Wajo is exported to Sabah. Pare-Pare, besides its role as a trading harbor, is also an important embarkation port for migrant workers en route to work in East Malaysia, particularly Sabah. The labor migrants mostly come from South Sulawesi and East Nusa Tenggara, particularly the Florenese from the East Flores District.

A study conducted by Parta (1996) in the central *Citra Niaga* market in downtown Samarinda shows that there is no ethnic dominance among shop owners. Ethnicity is a differentiating factor in the kinds of commodities sold in the shops. For instance, Banjarese sell jewelry and souvenirs, Javanese sell cooked foods in their *warung* (small restaurant), and Bugis usually sell cloth, shoes, electronic appliances, and fruit. The Dayaks traders often sell ethnic souvenirs, often identifying themselves as Kutais. In the mid-1980s, observers admitted the difficulties involved in estimating the population of Samarinda according to ethnicity: "From my casual observation in the city and from reading 1980 provincial-level census data, I argue that the last decade has seen a rapid increase in the number of migrants from other islands in the Indonesian archipelago. An important trend is the growing percentage of Javanese, particularly from East Java, moving into the city. An older but still significant migrant stream of Buginese from South Sulawesi is also helping to shape the ethnic composition of the city" (Wood, 1985, 131–132).

While a reliable source on the ethnic background of the dominant group in East Kalimantan is lacking, partial studies, among others by Magenda (1989), show that the Bugis are dominant, along with the Javanese, in the local bureaucracy in East Kalimantan, especially in the provincial government. The Bugis are also believed to be the dominant group in economic sectors, particularly in trading and transportation. According to Peluso (1987), the Bugis are an ethnic group in East Kalimantan that always assumes a crucial role when conflict or war occurs between ethnic groups. For example, in colonial times, the Bugis played an important role in defending the Kutais from the attack by the hinterland Dayak. The Bugis also defended the Kutai sultanate from the invasion of the Dutch and the Zulus, who came from the eastern sea (Mindanao, the Philippines).

Bugis Community in Samarinda Seberang

Samarinda Seberang is believed by the people in Samarinda to be the oldest Bugis settlement in Samarinda and perhaps in East Kalimantan. As noted by Pelras (1994, 321), the Bugis have settled in Samarinda since the early eighteen century.

In the early eighteenth century, the Wajo' prince La Ma'dukelleng, Arung of Sengkang (Sengkang-pen) and a fierce opponent of the Dutch and their Bone allies, left South Sulawesi with a group of about 3000 followers for Pasir in East Borneo, just across the Makassar Strait from South Sulawesi, where a small community of Bugis traders from Wajo' was already established. Engaging in trade, he stayed there until 1737, when he returned to South Sulawesi as the newly elected Arung matoa of Wajo' to wage war against Bone and its Dutch allies. In East Borneo, La Ma'dukelleng concluded political matrimonial alliances with local rulers just as he would have done in his home country; thus, he married one of his sons to one of the Sultan of Pasir's daughters and, later, their daughter to Sultan Idris of Kutei (Kutai-pen), while he himself became Sultan of Pasir. He was also active in organizing further Wajo' communities all down the east coast of Borneo, each under its elected chiefs.

Having obtained from the previous Sultan of Kutei the right of settlement for Bugis in Samarinda, a strategic settlement near the mouth of the Mahakam river, somewhat downriver from the Kutei capital, he later also obtained from Sultan Idris monopoly rights over the export products from the hinterland, including gold dust, benzoin, camphor, damar, gaharu wood, rattan, bird's nests, beeswax, bezoar stones, and rhinoceros horn (for which, however, only Kutei Malays were permitted to trade upriver), and sea products such as tortoise shell, turtles' eggs, agar-agar, and trepang. The Samarinda Bugis also had the monopoly imports of rice, salt, spices, coffee, tobacco, opium, chinaware, textiles, iron, firearms, saltpeter, and slaves. Moreover, their right to self-government was recognized, and they organized themselves under the authority of a chief entitled Pua' Ado (the daughter of the first bearer of this title had married another of La Ma'dukelleng's sons) and a council of a number of nakhoda or rich sea traders. Some Bugis leaders were granted titles by the Sultan, which put them for intermarriage with the Malay nobility and qualified them for intermarriage with the ruling dynasty. The Samarinda Bugis eventually obtained the right to control the upriver trade, and a Bugis was appointed shahbandar, an officer whose main function was to levy port taxes.

Pelras in his book, *The Bugis*, shows that the pattern and the process of migration that eventually formed a Bugis settlement differ from one settlement to another. The migration process of the Bugis in Samarinda, as shown above, indicates that political and economic changes both in South Sulawesi and in the land beyond play an important role in determining the existence and development of their settlements. Economy certainly plays an important role, and migration also can be seen as a Bugis market economy strategy. This is supported by Acciaioli (1998), who studied in the surrounding areas of Lindu Lake in Central Sulawesi how effectively the Bugis adjust their economic strategy of the market and price fluctuation of the agriculture products for exports.

Samarinda Seberang is an old Bugis settlement located almost in the center of Samarinda, the provincial capital of East Kalimantan. This area comprises two villages, namely Baka and Masjid. This Bugis community in Samarinda Seberang is particularly well known for its *sarung* weaving industry, which, interestingly, is known as Sarung Samarinda (not Sarung Bugis). It is a home industry employing mostly women. A small sample survey of 145 households, conducted in July–August 1999, shows that less than half of the persons in the survey are born in South Sulawesi (mostly from Wajo District). The other half are mostly born in Samarinda—as the second or third generation of migrants.

This fact proves that the Bugis have established their settlement in the area for a relatively long time. The large proportion of the Bugis born in Samarinda Seberang also indicates that migration seems no longer to be important phenomenon for the Bugis in Samarinda Seberang. The survey also reveals that the migrants, who migrated from South Sulawesi, ranged in age from 1 to 30 years. Perhaps permanent migration will decline, and circulation will become the new pattern of Bugis migration, from South Sulawesi to Samarinda Seberang. The improvement of sea transportation from South Sulawesi definitely plays a significant role in changing the pattern of Bugis migration to Samarinda Seberang from permanent to circular migration.

In Samarinda Seberang where Bugis culture dominates, a change also is apparently occurring, as seen by the findings of the survey. For instance, among the women workers in the Sarung industry, 103 were born in Samarinda, while only 76 were born in Wajo and 19 from other places in South Sulawesi. It shows that the ability and the skill to weave Sarung have already been transferred to Samarinda. The weaving skill, as part of the Bugis culture, is indeed a social–cultural capital that has no longer the monopoly of Wajo but has been preserved and developed by Bugis migrants and passed on to the younger generation in Samarinda. If the Sarung industries in Samarinda demand more labor, as often happens in a given period, employers will hasten to recruit in South Sulawesi for temporary employment in Samarinda Seberang.

During the visit to Sabbangparu, a subdistrict in Wajo that is well known for its Sarung production, in August 1999, there were only three houses still involved in the Sarung business. Each house only operated one weaving tool. From the interviews, the grim fact was revealed that the monetary crisis had devastated the industry. The price of imported thread for Sarung had drastically increased to a point beyond the means of the owners of the Sarung industry.[5] In order to survive, the people in Wajo turned their Sarung industry into an agriculture business, planting cocoa, particularly for export. Everywhere, people were planting cocoa, and as a result, the Sarung industry is practically extinguished in Wajo. You must go to Samarinda Seberang in East Kalimantan now to find the Bugis expertise in weaving Sarung. Another factor that has also contributed to the decline of the Sarung Bugis home industry is the rapid expansion of large Sarung factories in Java, in Gresik (East Java), Pekalongan (Central Java), and Majalaya (West Java), since the government invited foreign investment in the textile industry.

The ease of movement between the ports of Pare-Pare and Samarinda makes both places a continuum of sorts. In this continuum, massive movements of the population occur in shorter and shorter time periods. It could be argued that this demographic phenomenon of population mobility has removed administrative and geographic boundaries between provinces.[6] As an indication, a man who runs a Sarung Samarinda business and now lives in Samarinda Seberang describes how convenient it is to move by saying he could be *"going from Samarinda in the morning and by tomorrow afternoon has arrived in Pare-Pare."* Population mobility in this context has become the bridge linking two areas that are geographically separated, but then as time goes by, culture is also flowing and slowly integrated into a larger Bugis cultural domain.

[5]In the statistical book *Kabupaten Wajo Dalam Angka 1998*, published by the District Level Bureau of Statistics, there is not a single weaving industry remaining in the Wajo District.

[6]Lineton (1975a, b, 1) also observes similar phenomena: "the sending and the receiving areas are seen to be parts of a single society, each of which influences and is in turn affected by the other." In the case of Bugis migration to Jambi, for instance, Lineton describes a place that "provides opportunities for economic and political advancement which are absent or limited in the Bugis homeland." He adds: "It is possible indeed that migration is necessary for the continued functioning of many of the institutions of Bugis society, in particular the system of social stratification and political domination by aristocratic elite" (ibid, p. 3).

Ethnic Relations and Conflict Potentials

As a province that attracts many migrants, East Kalimantan has long been of interest to scholars, particularly on the relationship between ethnic groups. Peluso (1987), for example, studies the relationship between the Kutais, the Dayaks, and the Bugis, in connection with the trade economy before the arrival of the Dutch. According to Lineton, in the seventeenth century, the Bugis had already been given a strategic role by the Sultan of Kutai to defend the sultanate from attack by the Dayak who live in the hinterlands. As stated by Peluso (1987, 11), *"The Sultan's troubles with the Dayaks may have motivated him to grant the Bugis settlement rights in the territory now called Samarinda, the site of his capital at the time, as well as giving them a trade monopoly on the product of the interior."* Furthermore, Lineton argued, *"In exchange for these 'rights' of settlement, Bugis trader-warriors were to protect the sultan from 'hostile outsiders', including headhunting groups living beyond the bounds of the sultanate and Tausug or rival Bugis groups from other kingdoms. To this end, the Bugis supplied the sultan with warriors for his army as well as the firearms to equip those armies."*

After independence, studies on ethnic relationships in East Kalimantan are conducted in connection with, among others, the transmigration program. A study by Fulcher (1983) shows that acculturation is not happening between the Javanese transmigrants and the local Dayaks. While Evers et al. (1988) also indicated those trading networks are not developing between the Javanese and the Kutais as each group still primarily live within their own trading networks. A study on local politics by Burhan Magenda (1989) shows that the Bugis have successfully penetrated the local bureaucracy.

In East Kalimantan, the Dayaks and the Kutais are known as the indigenous populations, while the Javanese, the Bugis, and the Banjarese are regarded as migrants. According to an informant, a Dayak intellectual at the University of Mulawarman, before the 1970s, the population of the Dayak and the Malay, was 40–60, but now it has become 50–50. He believes that the ethnic compositional change has resulted from a strengthening of Dayak identity because some of the Dayak who previously preferred to be called Malay are now reasserting themselves as Dayak.[7] The revitalizing of Dayak identity, according to him, is because the Dayakness has become a unifying identity marker in relation to others, namely the Malay, the Banjar, the Bugis, or the Madurese. Religion also plays an important role in the relationship between ethnic groups in East Kalimantan. He further explained that the strengthening of Dayak's identity is also instigated by the perception among the Dayaks that they constitute an underdeveloped and oppressed group. The depletion of natural resources, especially deforestation, significantly contributes to increasing tension between the locals and the migrants, as the declining economic

[7]The results of the 2000 population census, however, proved that his judgment was incorrect. The local Dayak group comprises less than 10%, while the Javanese comprise 30%, the Bugis 18%, the Banjar 14%, and the Kutai 9% (Biro Pusat Statistik 2002).

resources result in growing competition between ethnic groups—especially between the locals and the migrants.

It was after the ethnic conflict in West Kalimantan between the locals and the Madurese migrants in 1997 that a need arose in East Kalimantan to build forums or an association of different ethnic groups in order to monitor and to control the possibility of ethnic conflict. In January 26, 1999, an ethnic-based association forum was established: *Forum Komunikasi Persaudaraan Antar-Masyarakat Kalimantan Timur* (FPKMT), or Forum of Communication and Brotherhood for the People of East Kalimantan, by the elderly members of each ethnic groups in East Kalimantan.[8]

For the Dayak of East Kalimantan, the violent ethnic conflicts in West Kalimantan, and later in Central Kalimantan, between the locals (particularly the Dayak) and the Madurese migrants, have further strengthened their perception of being indigenous. This, for example, prompted the establishment of Dayak's organization, *Ikatan Keluarga Pedalaman* (The Brotherhood of Hinterland People), particularly among the educated Dayak. The Dayak network has linked Dayak members not only in Kalimantan (Indonesia) but also in Sarawak and Sabah. This observation, for instance, is supported by a newly emerging issue in several Dayak districts near the border with East Malaysia and is strongly related to the local response to the new regional autonomy laws. As has occurred in other provinces in Indonesia, the demand to form a new province is also growing in East Kalimantan. Following the establishment of new regencies (Nunukan and Malinau) and the new Kotamadya (Tarakan), a discourse on the need to develop a new province in the north of East Kalimantan began to circulate, having first been proposed by the Bupati of Berau, Bulungan, and Tarakan. These three local leaders argued that the new province was needed to accelerate the development in the north, particularly in the border areas that have traditionally been neglected by both the provincial and central governments.

While ethnicity and indigenousness have never used in the discourse, casual observers will easily recognize that these elements are among the main factors behind the movement—and this, in fact, is the political demand among the people in the north: to be recognized as the owner of the region and its resources. The movement should therefore be seen as part of the emergence of ethnic politics in Indonesia—a new political development that is moving in parallel with the current seemingly confusing implementation of the decentralization and regional autonomy laws. In East Kalimantan, this movement is still growing but apparently not as fast as expected.

[8]A young local NGO leader in Samarinda, however, argues that the formation of FPKMT represents only the aspirations of a few elites, but not those of the common people in East Kalimantan.

Conclusion

The Bugis, in Samarinda, and East Kalimantan, generally, have long established their settlement and, more importantly, have developed a strong Bugis cultural domain in the land beyond South Sulawesi. While their numbers, as reported by the population census in 2000, are only the second largest population after the Javanese, their long-established settlement has provided a strong basis for their significant penetration of local economy and politics. Resentments toward the migrants in East Kalimantan have emerged, particularly among the local Dayak who feel marginalized by the economic and political domination of the non-Dayak populations. Yet, as we have seen, this resentment has not yet erupted into the violent conflicts we have witnessed in West and Central Kalimantan. While a proper investigation should be conducted to explain the reasons for this, my observations at least suggest two factors that might be behind these phenomena.

The first is the difficulty for the Dayaks to mobilize their resentment into a wider movement, as their number is proportionally small compared to the migrants; their residence is concentrated in the hinterland districts, mostly in the northern part of the province. The second factor is related to the fact that, aside from their large number, the migrants' populations, especially the Bugis, the Banjar, and the Javanese, have penetrated deep into the local economy and political structure, and it would now be very difficult for the locals to campaign against these group in East Kalimantan.

New regional autonomy laws, however, provide a new political arena that can be used as a vehicle to mobilize the politics of belonging, in which ethnicity and territoriality may be easily manipulated to become a political vehicle to face external domination. The movement from the local elites in northern districts demanding a separate province—*Kaltara* Province (Provinsi Kalimantan Utara), can be seen as the manifestation of the new politics—the politics of identity-based on ethnicity. In the case of East Kalimantan, it is perhaps interesting to observe the regularly active role of Syaukani, a district head of Kutai Kartanegara—a rich natural resources area, a Kutais, who is apparently becoming the important local leader who strongly advocates real autonomy for the district government. The political influence of Syaukani, who is also chairing the Association of District Governments in Indonesia (Asosiasi Pemerintahan Kabupaten Seluruh Indonesia or APKASI), is threatening not only the role of the governor but also the role of the central government, as he and his organization has become the strongest opponent for the central government's idea to revise the current regional autonomy laws.

Chapter 11
The Javanese in Lampung, Stranger or Locals?

> The Javanese continue to be the Javanese but modified by the Sumatran environment.
> *(Wertheim 1959)*

Introduction

Indonesia is well known as a country that consists of more than 300 large and small ethnic groups. What is also unique is its geographic contour; an archipelago consists of various ethnic groups residing on widespread islands. The Javanese is the largest ethnic group constituting approximately 40% of the Indonesian population, many of which live on Java Island. Considering the decline of Javanese people's welfare and a perception of an unbalanced distribution between Java and other islands, the Dutch colonial government shipped the Javanese to other islands. Lampung was firstly the first destination in 1905. Since then, the Javanese have slowly turned Lampung into another Java, outside Java, as the policy to relocate people is continued by the Indonesian government. Moreover, many Javanese subsequently decided to move to Lampung voluntarily since there are both an abundant supply of land and well-established Javanese communities, many of which causing the newcomers feel at home.

Nevertheless, beneath the appearance of a smooth and peaceful migratory process, tensions and conflicts frequently happened between the migrants and the native communities. Luckily, the authoritarian and repressive nature of both colonial and postcolonial governments are able to suppress the local disputes. In the last probably 10 years, following the fall of the unjust Suharto administration, the situation changes as the new decentralization policy is introduced by the new democratic government that has granted autonomy to the district-level governments. In the last 5 years, the governors and the district heads have been elected directly by the people.

This paper is originally prepared for the 2013 SIEAS Research Cluster Conference, Southeast Asia vis-a-vis Estrangers in the Historical Perspective, Sogang University, Seoul, October 31–November 1, 2013.

R. Tirtosudarmo, *From Colonization to Nation-State*,
https://doi.org/10.1007/978-981-16-6437-3_11

As expected, the local populations reasserted their local identities and mobilize their claims as the sons of the soil (*putra daerah*), demanding the right to govern their territories. The Javanese, who have become the dominant ethnic group in Lampung, have adjusted themselves to the new social and political settings. The resurgence of political identities, which is brought by the so-called *Lampungers,* has a political intention viz. evoking an emotional tension between the Javanese people vis-a-vis the native inhabitants. Yet, it is interesting to note that the Javanese individuals seem unaffected by such politicking and apparently not interested in the politics of identity.

This chapter, written as the result of a fieldwork in Lampung in 2011 and the study of some relevant literatures, aims to provide an explanation of why the Javanese people do not seem facing significant problems by being settlers in Lampung. In contrast to such focus, the question of whether or not the Javanese are locals is irrelevant to this particular article. Divided into five parts, this article, right after the introduction, describes the historical development of the general Javanese people within the context of Indonesian patriotism in the second part of this article. Moreover, the third part explores the process of Javanese migration and the resistance of local communities to their encroachment in Lampung. Next, the fourth part strives to fill the gap between what some literature has claimed and the social realities be found during the fieldwork, especially with regard to the recent implementation of decentralization. Finally, concluding remarks are provided in the fifth part.

Distinguishing the Javanese from the Chinese: A Comparison of Two Major Ethnic Groups.

Who are the actual Javanese that we are talking about? In Indonesia, at least, when people in everyday conversation talk about the Javanese, it is assumed that we (Indonesians) know what we mean by the term. There is something similar in our minds while thinking about the Javanese. It is perhaps something similar when we (all of us) consider the Chinese. The Chinese, at the global level, and the Javanese at the Indonesian level postulate a similar situation viz. their number is huge, and their presence overwhelms us. Allen Chun (1996), a Taiwanese expert in Chinese studies, in a mocking article entitled "Fuck Chinese: On the Ambiguities of Ethnicity as Culture as Identity," compels the following sentences:

> *It is said that China is the oldest extant civilization in the world and that its population constitutes one-quarter of humanity. Something so well entrenched demographically, territorially, politically, and historically should be anything but an uncertain entity. It is easy, thus, to identify something called "Chinese culture and society." Its political presence in the modern world system is incontestable, and the amount of intellectual discourse devoted to the study of China continues to fill libraries. In short, there is much to suggest that the very idea of China is an unambiguous or unquestionable entity. But what is so unambiguous about China that makes it an unquestioned object of gazing? What is the nature of Chineseness, and who are the Chinese? Finally, who is really speaking here?*

Chun then shows how diverse the meaning of China and Chineseness is:

While the Western term for China appears to accentuate the unity of a civilization brought about by the Ch'in Empire, Chinese terms for China and Chinese, on the other hand, suggest other kinds of associations, some of which are historically or regionally specific. The term chung-kuo, China's rendition of itself as "the middle kingdom," has existed since antiquity, and the term chung-kuojen is commonly used nowadays to denote Chinese people who speak chung- kuo hua or some form of Chinese language. When Chinese wish to talk about themselves as a unified people belonging to a unified culture, however, they refer to themselves as "people of the Han (dynasty)" (han-jen), as belonging to a Han culture that originated in the region of the Han River. The process of sinicization is one of being Han-ized (han-hua), and the ethnic minorities within territorial China are likewise set apart as being non-Han. Southern Chinese, in contrast, typically those from Fukien and Kwang-tung who constitute the vast majority of "overseas Chinese" in places such as Southeast Asia, express their Chineseness by saying that they are "people of the T'ang (dynasty)" (t'ang-jen) who speak "T'ang language" (t'ang-hua) and have deep attachments toward a homeland called "the land of T'ang" (t'ang-shan). Perhaps not coincidentally, Chinatown is called "street of the T'ang people" (t'ang-jen chieh). Nonetheless, the historical metaphor cannot be carried too far. When speakers refer to t'ang-shan, it usually means the China of one's home village and not that of the imperial court; likewise, t'ang-hua simply means "Chinese," which, because it can refer to any Chinese, does not sit well with Mandarin speakers who claim to speak chung-wen.

The Javanese are definitely not as overwhelming as the Chinese, yet we could discover some similarities between them. The Javanese people, like the Chinese, are indeed not a single entity as they are diverse in terms of geography, history, language, economy, and politics. Indeed, in the Java Island itself, we easily find many blatant differences among the Javanese identities, such as between the central and the peripheral identities, between the coastal and the mountainous identities, and between the western and the eastern parts. We could imagine, therefore, such diversity would spread out as the Javanese people, like the Chinese at the global level, move and migrate to other places either within or beyond Indonesia. No wonder then that we find *Pujakesuma* (the acronym of Putra Jawa Kelahiran Sumatra or the Sumatran Javanese), the Suriname Javanese, and Japung (the abbreviation of Jawa Lampung/Lampungnese Java). Recently, I visited the Netherlands where I met a Javanese who was born in Suriname, the Dutch colony in South Africa, where we know the Dutch coercively sent Javanese laborers in the 1920s and 1930s to unjustly work in various sugar plantations. This Javanese–Suriname person proudly told me about himself, "Suriname is my home, Holland is my country, and Java is my soul."

In the contexts of Java and the Javaneseness, Clifford Geertz (1960, 7) noted in the beginning of his book *The Religion of Java*:

Java – which has been civilized longer than England; which over a period of more than fifteen hundred years has seen Indians, Arabs, Chinese, Portuguese, and Dutch come and go; and which has today one of the world's densest populations, highest development of the arts, and most intensive agricultures – is not easily characterized under a single label or easily pictured in terms of a dominant theme. It is particularly true that in describing the religion of such a complex civilization as the Javanese any simple unitary view is certain to be inadequate; and so I have tried in the following pages to show how much variation in ritual, contrast in belief, and conflict in values lie concealed behind the simple statement that Java is more than 90 per cent Moslem. If I have chosen, consequently, to accent the religious diversity in contemporary Java – my intention has not been to deny the underlying religious unity of its people or, beyond them, of the Indonesian people, generally, but to bring home the reality of the complexity, depth, and richness of their spiritual life.

Having discussed the Javanese identities, perhaps it is appropriate to say that the Javanese people, like the Chinese, share both singular and plural identities.

The Unmaking of the Javanese Nation

Javanese and Chinese people are likely to ignite controversies not because they share various cultures, but their continual, constructed, and reconstructed identities pave the way to frictions. If there are differences between the Chinese and the Javanese, it may presumably be related to the notion that China does not only refer to a nation but also refer to a state. In contrast, there is no such thing called the Javanese nation nor the Javanese state. As a consequence of such a notion, we are aware of the existence of what is called Chinese citizens, but none of whom are called the Javanese citizens. In the Chinese particular context, we also notice the rivalry between the ROC (Taiwan) and the PRC, the extent to which both are contesting for political legitimacy and authenticity as being the legitimate heirs of Chinese culture.

From Indonesian history, we learn that the idea of Javaneseness loses its political legitimacy right after the founding fathers, many of which comprised of the cross, ethnic, educated elites, vowed on October 28, 1928, in Batavia declaring that they are "one nation, one language, and one country" that is called Indonesia. Ever since then, such an idea of Javaneseness merely indicates a group of people that are more or less united by a similar language, traditions and customs, arts, and cultural heritage. All those characteristics more or less signify what perhaps could be defined as the Javanese culture. Java, as indicated by the culture, is always in the making process, but Java as a political identity is unfortunately shrinking. In the Javanese studies, perhaps Benedict Anderson (1972a, b) is the one and only expert who elaborates eloquently what he calls as "the idea of power in Javanese culture." Meanwhile, John Pemberton (1994) scrutinizes the idea of reinvention of Java in his book *On the Subject of Java.*

To compare with the Chinese experience, Sun Yat Sen and Mao Tse Tung constructed a Chinese patriotism that needed to be based on culture, but Sukarno and Hatta ground their thoughts on the modern idea of the nation, which is based on civic instead of ethnonationalism. Indonesian nationalism earned its legitimacy from the national struggle against colonialism and the shared experiences and ideals of what Anderson describes as the imagined community (2003). Moreover, Indonesian nationalism is not defined by the glories of the past but by the common goals on the future. Indonesia is a common political project that should be realized by the common efforts of all ethnic groups. The construction of nationhood is indeed influenced by the way nationalism is crafted. The state as the embodiment of the nation often requires a kind of national identity. National symbols in the forms of arts, languages, architectures, and food are inevitably necessary as the characteristics of national identity and nationhood. In this regard, it is interesting to remember that a nation always consists of different ethnicities that possess their own ethnic characteristics. In the case of China, despite the fact that the Han is the largest and most dominant ethnic group, there are minorities that also possess their ethnic identities. Such a situation poses a problem as Allen Chun concludes:

> *The existence of ethnic minorities within China's national borders, on the other hand, created particular problems for reconciling notions of ethnicity in relation to national identity. For the most part, the government adopted a Stalinist policy of multiculturalism, with the Han ethnic majority positioned at the core of a family of nations.*

Ethnically speaking, Indonesia is probably more diverse than China although the Javanese, like the Han ethnic, could be regarded as the major ethnic group. The Javanese people, however, do not have the luxury to claim as the core of a family of nations as the Han in China does. On the one hand, the Han is deliberately constructed by the state as the core nation. On the other hand, the Javanese have never been so well constructed. The construction of the Han as the core nation, however, implies some consequences as shown by Allen Chun:

> *The criteria of traditional practices and material customs applied to distinguish ethnic groups, however, made the very notion of ethnic identity within a cultural taxonomy problematic, especially in cases of historically known minority groups that had been undergoing a long process of sinicization. In this regard, the need to define ethnicity according to the kind of hard-and-fast rules characteristic of the boundedness of a modern nation-state ultimately fabricated ethnic divisions that did not exist in the minds and lives of the people themselves, while at the same time made cultural objectification a normative practice in the state's institutional routine. As in the case of Taiwan, history and ethnicity thus combined to produce (a national) identity in which they were, in fact, nothing more than imagined constructions by the state to define the ethos of its own modernity.*

A long process of sinicization as experienced by minority groups in China also occurred in Indonesia, culminating during the reign of the Suharto military government from the 1970s to the 1990s. The process of Javanization had already become a contentious issue in the 1950s at the time when the national government began advocating the transmigration programs as a policy to assimilate the local populations into the so-called national identity. Heeren (1979) and Wertheim (1989) observed the implementation of this policy in Lampung and Sumatra Selatan provinces. The national government created some enclaves of Javanese resettlements in the middle of Lampung communities. In Suharto's administration, the transmigration program is intensified by the support from the World Bank both for controlling the population density in Java and for supporting what they called as regional development. The ultimate goal of relocating the Javanese through transmigration programs was, however, ideological: to achieve the national integration (Tirtosudarmo, 2001).

The policy has been widely criticized because it implies both marginalizing the minority groups and deforestation in order to open a new land. It is no secret that the timber companies take benefits from the transmigration program. On the pretext of strengthening national integration, the transmigration program is also used in order to reinforce the strategic interests of military forces in several areas, perceived by the national government as a kind of fortified bastion against the separatist movements such as in Aceh, Papua, and East Timor. Despite all controversies concerning the security system of transmigration program during the Suharto administration, it is still relevant to ask whether or not the durable process of Javanization in Lampung as the oldest place for relocating the Javanese, and as a result, the province is by now demographically dominated by the Javanese, did come into reality.

Given that Java has essentially failed to be the basis for constructing the Javanese nation, in another paper (Tirtosudarmo, 2005, 14–15), I have explained the reason for such failures:

> *The explanation as to why the Javanese failed to assert their political identity should be sought in the history of nationalism in Indonesia which began to emerge in the dawn of the twentieth century. The first generation of Javanese intellectuals that were the product of the Dutch educational system began to imagine what sort of future political community would suit the indigenous people in the archipelago. Here I would like to cite the debate between two Javanese intellectuals, namely Tjipto Mangoenkoesomo - who advocated Indies nationalism - and Soetatmo Soeryokoesomo- who advocated Javanese nationalism. This debate – in Dutch not Javanese or Malay – took place in 1918, the same year in which the Volksraad (People's Council) was founded by the Dutch (see Shiraishi,* 1981*). In this debate, Soetatmo advocated Javanese nationalism, arguing that the nation could and should be built on the basis of common culture and language. Javanese nationalism had its basis in the common culture, language and history of the Javanese, whereas the cultural bases of Indies nationalism were nonexistent or, at best, a product of Dutch colonial rule. Javanese nationalism was the means of self-expression for the Javanese, while the Indies nationalism was no more than a reaction to Dutch colonial domination of the Indies. Therefore, he (Soetatmo) argued, only Javanese nationalism had the sound cultural basis on which the Javanese could establish their future political community. In reaction to this argument of Soetatmo, Tjipto defended Indies nationalism. In his opinion, what was totally lacking in Soetatmo's view was world historical development. He argued that Europe was clearly more advanced than Asia, and therefore the Javanese could learn from the European historical experience the direction in which the national formation in the Indies would go. The Indies were indeed composed of diverse ethnic groups, with each ethnic group having a different culture and language, but Java had lost its sovereignty and was only a part of the Dutch- dominated Indies. The fatherland of the Javanese was no longer Java but the Indies, and the task of the national leaders was to work for Indies nationalism. In the end, after a long process of negotiation and conflicts, Indies nationalism more or less prevailed as the new form of "Indonesian nationalism." Yet, Javanese-ness did not fade away, but instead contributed – in some instances through elite manipulation and reinvention of Java - to contemporary Indonesian politics (Pemberton* 1994*).*

Like China, Indonesia also faces some difficulties in the construction of national identity. National identity as perceived by the power holders is required by what they define as the national culture. But what is national culture? The constitution defines it as the peaks of all regional or local cultures. If this is the case, the next question is how do we opt the regional or local cultures that ultimately represent the national culture?

Javanese Migration and the Lampunger Resistance

The Javanese diaspora has existed since the earliest times as Hugo (1980) notes, "A good example is the Javanese movement into the Banten and north coastal regions of Sundanese West Java which gaining its momentum in the sixteenth century …". Furthermore, he argues that:

> *This movement was instrumental in the spread of sedentary wet rice cultivation into West Java where previously almost all of the indigenous Sundanese population engaged in ladang (dry field) cultivation.*

The presence of Javanese people in Sunda and Banten has become, in turn, the source of migration to Lampung.[1] According to the late Indonesian maritime historian Adri Lapian, the Sunda Strait is geographically decisive in determining the interaction between people in both Banten and Lampung. Thereby, discussing Lampung should be necessarily complemented with Banten according to Lapian.[2]

In 1905, the unjust Dutch colonial government began to relocate people from Java to Lampung. Under the so-called ethical policy, relocating people (land colonization) was one out of three related programs, with the other two are education and irrigation. Through the land colonization program, 155 Javanese families from Bagelen, Kedu areas in Central Java, are resettled to Gedong Tataan in Lampung, Southern Sumatra Selatan.[3] The motivation of the Dutch is not merely the demography and economy-based reasons but also due to security reasons. The Javanese countryside in the nineteenth and early twentieth centuries is characterized by sporadic social unrest and the rebellious peasant movements against the Dutch unjust, repressive agricultural policies (Kartodirdjo 1973). Not only did sporadic, the peasants' unrests and violent conflicts occurred almost on the entire Javanese rural areas between 1900 and 1920 in Tangerang, Pamanukan, Sukabumi, Ciasem, Kuningan (Jawa Barat), Pekalongan, Gombong, Semarang (Jawa Tengah), Mojokerto, Sidoarjo, Kediri, and Jember (Jawa Timur) (ANRI 1981). There might be a direct linkage of the Javanese rural unrest and land colonization in Sumatra, although it is inconceivable that such a coincidence could posit a causal relationship.

The perception that land colonization is not only driven by some demographic reasons but also aimed to be a kind of social experimentation that is also conveyed by ORSTROM (1986) in their report:

> *Lampung's destiny has been a prodigious one. Devoid of human inhabitants up to the end of the 19th century, this southernmost tip of Sumatra was to become, in the 20th, the privileged experimental zone for all the forms of organized colonization ever elaborated in the*

[1] There are at least two theories about the place of origin of people in Lampung. The first theory holds that their ancestors came from Banten in the western part of Java, while in contrast, the second explains that the origin of the people in Lampung is Pagar Ruyung in Sumatra Barat.

[2] Personal communication in May 2011 with A.B. Lapian in Jakarta.

[3] Gedong Tataan is located at the foot of Mt. Betung, about 20 km west of Tanjung Karang (Bandar Lampung). At the time, Gedong Tataan became a subdistrict of the district of Pasawaran, and Bagelen became the name of a village in the Subdistrict Gedong Tataan.

framework of the Dutch kolonisatie or the Transmigration program of independent Indonesia. The voluntary development policies were so successful there that today the government has been forced to take preventive measures in order to protect the natural environment and to rely upon the local offices of the Transmigration Ministry to encourage emigration to the other islands in the archipelago.

The Dutch land colonization program that is perceived by ORSTROM as a social experimentation experienced several changes since its beginning in 1905 until the Javanese arrival in 1942 (ORSTOM 1986, 83–87). Heijting, the architect of the land colonization program, says that "Javanese emigration can only be successful by establishing 'little Java' in Lampung."[4] The conceptual idea on which the land colonization program is based—known as the "Javanese paradigm"—in principle is aimed to create a Javanese rural community in Lampung. The Javanese are relocated into an area that is considered by the Dutch as an empty land in the middle of local communities in Lampung, as noted by Kingston (1990):

In recreating the Javanese landscape, the government allotted about 1.5 bouw (1 bouw= 0.709 ha.) of land per family and encouraged wet rice cultivation by subsidizing irrigation. The government intended that the pioneer settlements engage in food production and did not encourage export commodity crop production; the Dutch image of what Javanese villages ought to be like overshadowed what they could be like. It was the ideal type of Javanese village, conveyed by the village elite and embraced by the colonial officials that defined the manner in which transmigration was implemented on the ground.

Heeren (1979) argues that the reason for Heijting to relocate the Javanese in an area that is separated from the local communities was not primarily to make the Javanese people feel at home but due to the resistance from local communities in handing over their land to the Dutch unjust colonial government. Separating the Javanese communities from the local population, however, implied criticism from Nitisastro (1955, 125, as quoted in Heeren 1979, 17) by which such policy reflects the enclaved politics. Widjojo Nitisastro is of the opinion that the Javanese should be able to assimilate with the local population since both communities are Indonesian citizens. Based on his observation in 1956 in Lampung, Wertheim (1959) finds that assimilation between the Javanese and the local population occurred in practice, but it did not turn the Javanese into Sumatran people or Indonesians, but rather be more Javanese with some adjustments to the Sumatran environment.

Kampto Utomo (1975) notes that until 1928 the resettlement pattern of the Javanese indeed followed the enclaved politics where the Javanese are engineered to form their own local governance.[5] Until the *marga* system is revived by the Dutch as "*inlandse gemeente,*" the Javanese communities are governed under the *marga* system. The Javanese communities spread out very fast ranged from Gedong Tataan to the west, and occupying places such as Pagelaran, and to the south encroaching the areas surrounding Rawa Kementara. In the western part, starting from Pringsewu

[4]See Kingston, J. 1990. Agricultural Involution among Lampung's Javanese?. Southeast Asian Studies, Vol. 27, No.4, March.

[5]Kampto Utomo's study in Lampung was the basis for his PhD thesis at Bogor Institute of Agriculture where Professor W.F. Wertheim was his supervisor.

that later on transform into a Javanese concentration, and to the north encroaching areas of Way Sekampung. The encroachment of Javanese in Lampung practically halted during the Japanese occupation but resumed immediately after the Japanese left (1942) and the recognition of Indonesian sovereignty by the Dutch in 1949. According to Kampto Utomo (1975), the reason for resuming the land encroachment by the Javanese in Lampung was primarily due to the communication and interaction between themselves with their relatives in Java run smoothly as people visited one another or sent letters that eventually encouraging the unstoppable, voluntary transmigration.

The waves of Javanese resettlement in Lampung, as Wertheim observed in 1956, eventually created some social tensions with the local population, in which case the land competition inevitably increased:

> *This situation leads to an increasing resistance of Sumatrans to the way in which resettlement has been carried on. Such resistance may seriously hamper further transmigration efforts. Thus it can be stated that the absorptive capacity of the outer islands is not only restricted by spatial and technical but also by social factors as well.*

Who are actually the local communities in Lampung resisting such Javanese encroachment? Referring to the definition used in the 2000 Population Census, those are Orang Peminggir (6.42%), Orang Pepadun (4.22%), and Orang Abung Bunga Mayang (1.28%) (Suryadinata et al. 2003, 18). In their research on the local communities in Lampung, ORSTROM (1986, 91) identifies at least four groups: Orang Abung, Orang Menggala Tulang Bawang, Orang Pubian, and Orang Pesisir. The migrant groups originating from South Sumatra, as identified by ORSTROM, are Orang Ogan, Orang Sumendo, Orang Way Kanan, and Orang Mesuji. In relation to the Javanese settlers, these people could be categorized as the local population or simply called Orang Lampung or the Lampungers.

Wertheim's observation in 1956 is relevant to the wider political context after the Dutch left Indonesia. After the independence, the administrative system at the local level began to change in 1950 following the rapid change in the political system at the national level. The change in the land ownership system (previously organized through the *adat* (customary) laws but then the national laws) is one of the sources of dissatisfaction among the local population. The reorganization of *marga* (from the local government's hierarchical structure into the new autonomous institution called *negeri*) also triggered social tensions. The *negeri* comprises several *marga,* including the Javanese communities that reside into the respective areas (Heeren, 1979). According to the national laws, a *negeri* should be created based on democratic principles rather than by the customary laws. Therefore, a Javanese could be the head of a *negeri* insofar as he/she is elected democratically.

The various changes related to the new arrangement concerning local governance caused local grievances against the national government. Both the regional rebellions in Sumatra Barat, Sulawesi Selatan, and Sulawesi Utara under the banner of PRRI/Permesta in 1956–1958 paved the way for local populations in Lampung to articulate their feelings through a petition to stop the transmigration program, which is launched in the *Adat* Conference, firstly in Palembang (January 1956) and then in

Bukittinggi (March 1957).[6] In response to the threatening political situation in 1959, President Sukarno, supported by the military forces, announced a decree to return to the 1945 Constitution and arbitrarily dissolved the Konstituante that was preparing a new constitution. Under this administration, the *negeri* and the *marga* systems are abolished as well, and all local administrative arrangements are organized by the district governments.

As the regional rebellions finally reduced the frequent resistance of the local population in Lampung against the Javanese settler, transmigration is also practically halted due to the budget constraint of the national government. Indeed, the program is stopped due to the arbitrary change in power from Sukarno to Suharto following the 1965 national political tragedy. In the 1970s, the transmigration program is re-implemented under Suharto's New Order regime, in which Lampung was one of the major transmigration destination areas.[7] Here, the nationalization project of local government is reshaped by Law No. 4, 1974, concerning the regional-level government, and Law No. 5, 1979, concerning the village-level government. The uniformity process of local government throughout Indonesia proceeded almost without any significant resistance from the local population as Suharto arbitrarily appointed military officers to be both the district heads and the provincial governors.[8]

The Javanese in Lampung After the New Order

The step-down of Suharto, a Javanese, from power in May 1988 is described by a foreign observer as the end of the "Last Sultan" in Indonesia (Loveard 1999). Nonetheless, Suharto, who held almost absolute power for more than 30 years, is not considered as the last Javanese king. Although Suharto's style of government is strongly associated with the Javanese political culture (Pemberton 1994), only the Yogyakarta's Sultan is popularly known as the Javanese king. The collapse of Suharto's authoritarian political regime is perceived as the end of a centralist government that emphasized too much uniformity in the arrangement of local government. It is, therefore, quite understandable that one of the immediate demands in the aftermath of Suharto regime should be for political decentralization and regional autonomy.

[6]According to Heeren (1979, 50), the abolition of the transmigration program was also one of the demands of PRRI/Permesta.

[7]The following is the national number of families moved under transmigration programs and the percentage. Repelita I (1969–1974): 11397 KK (29%), Repelita II (1974–1979): 4500 KK (8%), Repelita III (1979–1984): 42876 KK (12%), Repelita IV: 1984–1989: 17893 KK (9%), Repelita V (1989–1994): 12515 (5%), Repelita VI (1994–1997): 8412 (4%). Tirtosudarmo (2001, 212).

[8]In Lampung from 1973 to 1998, the governors are always a Javanese military general: Sutiyoso (1973–1978), Yasir Hadibroto (1978–1988), and Pedjono Pranyoto (1988–1998).

Several new regional laws and regulations are created to support the implementation of a decentralization policy to open up the political space for the local populations to articulate their wishes. Under the new political space, local populations began to mobilize their local identities to reclaim what they conceived as their local rights. *Adat* or customary laws that are totally repressed under Suharto administration is revived by now. The local populations reclaimed their land ownership and organized their right to have their own local government and leaders based on the *adat* laws. In Lampung, local intellectuals actively reinvent their traditions in order to provide the justifications for their political claims, such as the sons of the soil (*putra daerah*). They have reinvented traditions—to borrow Hobsbawm's terminology—through several projects such as rewriting the history of Lampung, reviving local customs, and teaching the Lampung language. Such activities, however, have implied limited impacts and are still in the early stages.

Migration of the Javanese to Lampung has occurred for more than a century so that Lampung has become demographically and culturally a kind of Javanese Province outside Java. In many places such as Pringsewu and Metro, for example, Javanese has become the lingua franca. In this regard, the question put forward in the title of this paper "The Javanese.... locals or strangers?", seems to be relevant here. As the ORSTROM (1986) study has shown, the majority of Javanese in Lampung comprises the second, the third even the fourth Javanese generations. These "Javanese–Lampung" as they are called do not even know where their Javanese ancestors originally came from. The Javanese have dominated various positions and occupations, within the government offices, in the universities and local parliaments, as well as in the private sectors. As the position of the governors and the district heads is now contested through direct elections, the Javanese should be quantitatively the winners if they politically exploit their ethnic identities to win the elections. In contrast to such political configuration, the Lampungers would not mathematically vanquish the Javanese in elections though they could exploit their claim as the sons of the soil.

The new electoral system allowing the direct election of local leaders paves the way for the local population to mobilize their local identities to influence the election processes. In such a situation, Lampung provides an interesting case as the majority of its population is Javanese, and the local population is a minority. In the last gubernatorial election, the candidates used cultural symbols to attract votes. Interestingly, the Javanese candidates are lost to the local Lampunger candidate. This reality could be interpreted as that the Javanese voters—that is, the majority—apparently are not attracted to the Javanese candidate, and in fact they chose the local candidate as their governor. How can we explain this phenomenon?

According to Wertheim's (1959) observations on Lampung in 1956 says, the Javanese identity developed not according to the expectations of the designers of land colonization or transmigration program. Both the designer who wanted the Javanese to retain their Javanesness by engineering the enclaved politics and other designers who expected the Javanese to be the agents of assimilation to create an Indonesianess in Lampung should have disappointed by the reality as the Javanese maintain their Javaneseness but be modified into the Sumatran environment to some extent. To be consistent with Wertheim's study, the Javanese in Lampung turn out to be the "Jawa–Lampung" or "Japung." This new identity of Javanese–Lampungers is a result of more than a century of Javanese cultural transplantation into the Lampung environment. Indeed, the Javanese identity in Lampung is so much different from the Javanese in Solo or Yogyakarta, the Javanese on the northern coast of Java, or the Javanese in Suriname or New Caledonia.

The political participation of the Javanese in Lampung, as reflected by their decision to vote for the local candidate instead of the Javanese candidates, indicates their pragmatism in their daily lives. For the Javanese, it is not important what the ethnic identity of the candidate is insofar as he or she is perceived as a capable leader to foster peace and welfare to the society. The current euphoria in many communities to create a new district in Lampung, for instance, is also not a big deal for the Javanese as they might think that whatever happens, they are the majority in any way.

Conclusion

Javanesness, unlike Chineseness, has not turned into a political identity. Some attempts done by a few Javanese elite groups to form a Javanese nationalism are proven to be unsuccessful because Indonesian leaders deliberately developed a civic instead of an ethnic-based nationalism (ethnonationalism). The Javanese have submitted their ethnic identity in favor of the trans-ethnic and national identity though they are demographically the dominant ethnic group in Indonesia. In other words, Javanesness transformed into merely a social and cultural identity rather than political identities. In the era of the post-Suharto political decentralization, as the case in Lampung has shown, the Javanese seem not to be interested in mobilizing their ethnic identity as it has been excessively taken for granted by many other ethnic groups in Indonesia.

Apart from the historical fact that the Javanese have relegated the idea of Javanese nationalism, in reality, they have already gained some economic and political important positions in Lampung. Their votes are always decisive to the local election results. In Lampung, the issue of ethnicity is an elite issue, while in contrast, laymen, either Javanese or non-Javanese, pay no attention to such topic. Tradition is exercised by the local elites because they foresee some political opportunities provided by the new political developments in the post-Suharto administration. The discourse of Lampungers as the sons of the soil (*putra daerah*) has increased in the last ten years, and indeed, there is a common question to the Javanese, "how should they respond to such increasing politics of identity in Lampung?" Based on my short fieldwork in Lampung, I have come to several arguably premises regarding the main questions put forward in the title of this paper, "The Javanese in Lampung: Strangers or Locals?"

Firstly, the Javanese in Lampung is not really concerned about their ethnic identity. Secondly, if tensions and conflict occur in such a way, the source is not really ethnicity but rather the rivalry over various economic resources, especially the land ownership. Thirdly, Javanesness apparently could not be mobilized for supporting the politics of identity in Lampung. Fourthly, the Javanese and Javanesness are certainly not static and homogeneous but diverse and constantly shifting in response to the changing environment. Fifthly, the Javanese in Lampung have created a new identity viz. the "Javanese–Lampung" or "Japung" that could be considered as the local identity. Yet, such thing is no longer relevant because they are somehow regarded as strangers.

Chapter 12
Disintegration from Within? Ethnicity and Decentralization Politics

Indonesia, in contrast with its closest neighboring state Malaysia, is born as a nation with a strong sentiment of civic nationalism. The Javanese, although demographically is dominant ethnic group, accepted non-ethnic-based nationalism. The Malays, claiming as the "son of the soil" in Malaysia, advocated strongly that "Malayness" should be the fundamental basis of the state's ideology and nationalism.[1] At the beginning of the twentieth century, in a place that later becomes Indonesia, ethnicity, religion, and place of origin are important binding factors in the creation of youth and student organizations that were actively taking part in the independence movement. Interestingly, these "primordial"-based organizations have not become significant divisive forces in the construction of strong Indonesia's trans-ethnic nationalism. On October 28, 1928, leaders of youth organization congregated in Batavia and declared an oath "Sumpah Pemuda" as "One Nation, One Language, and One Land"—"Indonesia." "Indonesia" an obscure Greek's word is firstly coined by G.W. Earl and J.R. Logan in their articles written in 1850, which is then used by a progressive student organization in the Netherlands and become the important symbol of nationalism and the rallying slogan for the nationalist leaders in the independence movement.[2] On August 17, 1945, Indonesia proclaimed its independence. Secular ideologies (nationalism, communism, and socialism) apart from Islam then become the strong basis for party politics after independence. Ten years after the declaration of independence, the first general election was conducted in 1955 in which the votes are evenly divided into the three largest political parties: communist party, nationalist party, and Islamic parties.

[1] I have elaborated the distinction and interaction between the Javanese and the Malay in the process of postcolonial nation building in the CRISE Working Paper No. 14, "The Orang Melayu and Orang Jawa in the 'Lands Below the Winds': Notes on the Historical Imprints of the 'Civic-Ethnic' Distinction in Indonesia and Malaysia".

[2] See article by Jan B. Ave "'Indonesia', 'Insulinde' and 'Nusantara': Dotting the I's and Crossing the T's." BKI.

R. Tirtosudarmo, *From Colonization to Nation-State*,
https://doi.org/10.1007/978-981-16-6437-3_12

Until 1965, Indonesian politics is basically dominated by three political mainstreams apart from the increasingly important balancing political role played out by President Sukarno and the military. American Anthropologist Clifford Geertz conducted fieldwork in Java during the early fifties and continue until the early sixties introducing the sociopolitical concept of *aliran* or stream as the underlying cultural element in Indonesian politics: *abangan, priyayi, and santri.*[3] These three cultural identifications resemble the mainstream politics during that time: *abangan* associated with the communist party, *priyayi* with the nationalist party, and *santri* with Islamic parties. After communist was crushed in 1965 and Sukarno was forced out from politics, only the military and the weak nationalist and Islamic parties left in the Indonesian political landscape. The military under General Suharto then become the second president, slowly transforming itself into a powerful organization that since then has infiltrated the national politics and civilian government. The Indonesian political terrain was "depoliticized" under the new quasi-ideology of "developmentalism" that is supported by the US and Western countries to focus its national development strategy on economic growth. With the fall of Suharto from power in 1998, politics is revived, and political parties flourish under the new slogan of *reformasi* (political reform) and democratization. In this situation, the constitution was amended, the general election was freer, and overcentralization was loosening up. It is within this changing climate of decentralizing politics that ethnicity has emerged as the strong mobilizing cultural identities in the post-Suharto political landscape. The emergence of ethnic politics in the current decentralizing era could have detrimental impacts on the nation building that is based on the idea of pluralism and civic nationalism that brought Indonesia into being. The strengthening of ethnocentrism that is developed in conjunction with the implementation of new decentralization policies could be seen as a process of political disintegration from within.

Politics of Decentralization in Perspectives

This chapter is a commentary on the emergence of ethnicity in the current political development as manifested in decentralization politics. Several studies have shown that ethnicity apart from religion is implicated as the crucial elements that are played out in several communal conflicts that occurred recently. How the political system after the fall of Suharto in 1998 could prevent the violent conflict emanating from the existence of horizontal inequality—an inequality between different culturally defined groups? Or put the question differently how decentralization politics contribute to the emergence of ethnic politics? This chapter will limit the discussion on the issue of ethnicity and conflict and how it relates to the decentralization politics in the post-Suharto era. This chapter excludes violent conflict involving anti-Chinese

[3]See Clifford Geertz's. 1976. "*Religion* of Java." Chicago: The University of Chicago Press.

riots, military attacks on the Islamic groups (East Java, Lampung, Tanjung Priok), and other non-ethno-communal conflicts. While conflicts in Maluku, Kalimantan, and Sulawesi could be labeled as communal, the conflict in Aceh and West Papua could perhaps be labeled as political. Apart from the inequalities between the core and the peripheries, both conflicts are also closely related to the economic and political inequalities between different culturally and politically defined groups.

The decentralization policy is attempted by the successive post-Suharto governments to ease the local grievances toward the previous centralistic development strategy that is resulted in the political injustices and the growing inequalities between the core regions and the peripheries. Decentralization is not a new concept in Indonesia. The center and regional relation has always been an important political agenda and constantly being negotiated throughout the life of the republic. The national government has become more and more centralistic after the failure of regional rebellions in 1956–1958 that eventually elevating President Sukarno and the military in the center of national power. After the 1965 bloodiest political tragedy of the communist killings, it has become apparent that only the military was left in the center of power. Furthermore, under Suharto, political parties were marginalized, and the regions become peripheralized. In this period, decentralization is conducted within the contexts of overcentralization and the deepening process of mystification of Indonesia as a conclusive and final unitary state. The process of mystification of Indonesia as a unitary state—*Negara Kesatuan Republik Indonesia* or *NKRI*—is an important political development that has long-term consequences on the way the discourse and practice of nation-state building processes are taking place and in turn on the way decentralization politics is presently carried out.

The mystification of *NKRI* began with the rising of Suharto as the sole power and the "militarization" of Indonesian politics. The military's strong perception of the nation as essentially constituting a spatial entity with clear geographic boundaries that should be jealously guarded from external threats has become the underlying factor in mystifying *NKRI* as a sacred cow with all its taboo that prevents any critics to its supposedly principal substance. Constructing *NKRI* as a national myth for the military is a strategic way to sustain the centralistic political control of the vested interest groups, particularly the military.[4] The discourse on national integration become the monopoly of the military and was continuously propagated by its

[4]It is remarkably interesting to observe how the term *NKRI* is used by military leaders. The Indonesian military is well known for its habit of using abbreviations in their official language. The common usage of *NKRI* by the military in their language therefore appears to be a normal practice. Yet, in the case of *NKRI* it should be observed more critically as the term apparently has been elevated into a "security mantra" to suppress any perceived threats to national integration and state's sovereignty. In the commemoration of "Hari Kesaktian Pancasila" in the October 1, 2005, the Head of Parliament, Agung Laksono, reads the oath that says, among others, "*sejak Proklamasi 17 Agustus 1945 terjadi banyak rongrongan dari luar negeri terhadap NKRI. Rongrongan itu antara lain berupa upaya menumbangkan Pancasila, namun karena kebersamaan yang dilandasi nilai-nilai luhur ideologi Pancasila, bangsa Indonesia dapat mempertahankan tegaknya NKRI.*" "*Maka, di hadapan Tuhan YME dengan memperingati Hari Kesaktian Pancasila kami membulatkan tekad untuk tetap mempertahankan dan mengamalkan nilai-nilai Pancasila, sebagai*

military's think tank—the *Lemhanas* (National Defense Institute). The internalization of military dominating discourse on national integration into civilian bureaucratic officials is systematically conducted through the compulsory courses of public servants regularly conducted at *LAN* (National Administrative Institute). In the New Order's period, a narrow version of Indonesian nationalism has entrenched and become the source of intimidation and persecution toward the government's critics and any groups that are perceived as a threat to the national integration.[5] From this narrow nationalistic viewpoint, the discussion on federalism as a logical political alternative for Indonesia is understandably out of the question.

The mystification of NKRI and the dominant military's discourse on national integration and its related national security threats constitute the broader framework in which the conceptualization and practice of decentralization and regional autonomy are taking place during Suharto and the post-Suharto periods. It was during the New Order that the idea of providing autonomy to the district-level government and not to the province began to be implemented. In 1992, Suharto's government provided a regulation (Peraturan Pemerintah No. 45) that is aimed to implement the regional autonomy that is given to the district-level government. It was then, under the retired general Rudini as the minister of Home Affairs, that a project on regional autonomy was piloted in one selected district in all provinces. The rationale to give autonomy to the district-level government is based on the National Law of the Arrangement of Regional Government No. 5, 1974, in which public services should be provided by the local government that close to the people—namely the district-level government. The power and authority provided to the regional government during Suharto's New Order, however, very limited and basically aimed to ensure the implementation of central government development plans and programs in the regions. The regional government is given autonomy only to faster the implementation of central government development plans. In other words, the decentralization policy under Suharto is merely an extension of its centralistic national policy.

In the post-Suharto period, new enthusiasms are clearly expressed to revoke the New Order's decentralization concepts. Decentralization and regional autonomy have quickly become one of the important national agendas under the banner and the new slogan of posting Suharto politics of *reformasi* (political reform). It was under the seemingly rapid political change that the idea of transferring power from the center to the region has been closely negotiated among the political elites. Teamwork—headed by Ryaas Rasyid—was established under President Habibie's

sumber kekuatan menggalang kebersamaan untuk memperjuangkan menegakkan kebenaran dan keadilan demi keutuhan NKRI." Quoted from Media Indonesia, October 2, 2005, page 12.

[5]It is not surprising therefore that the issues of secessionists' movements—in East Timor, West Papua, and Aceh—received more publicities after the military labeled them as *GPK* (Gerakan Pengacau Keamanan) and *GPL* (Gerakan Pengacau Liar)—simply means "trouble makers." These labels were introduced by the military to create the image that these rebel groups are merely criminals and not freedom fighters. For related discussion on this issue, see, for example, Dave McRae's "A Discourse on Separatist," *Indonesia,* 74 (October 2002) pp. 37–58.

auspices in 1999 to draft the new regional autonomy law. The failure of the Habibie government in stopping the East Timor independence in August 1999 provides a strong political precedence to lay the legal and structural foundations for the state to balance the power relationship between the central and regional governments. In 1999, two important national laws dealing with decentralization and the position of regional government vs. the national government were produced in 1999: Law No. 22 on "Regional Government" and Law No. 25 on "Financial Balance between Central and Regional Government." These two national laws replaced the previous national laws that arrange the regional government—Law No. 5, 1974, on regional government and Law No. 5, 1979, on village government.

Now, 5 years after its implementation in 2001, the new regional autonomy has generated high political expectation and proliferate short-term economic demands that go beyond the capacity of the central government to accommodate. While the impact of the new decentralization policy cannot be generalized, it was become apparent, especially during Megawati presidency (2002–2004), that the central government perceives the danger of possible conflicting authorities in the provincial government as the laws contain several weaknesses in delineating distribution of power between provincial- and district-level governments. Excessive interpretation of the notion of autonomy by the local leaders at the district-level government has resulted in the uncontrollable exploitation of natural resources. Another contention is related to the regulation for electing the head of the regional government in which for the first time will be conducted through direct election. The autonomy syndrome easily slips into the idea that the region should belong to the locals. The "son of the soil" or *putra daerah* should be given priority in the recruitment of public servants and in local governmental posts that are clearly very discriminative for the nonlocals. *Daerahisme* or "localism" is, in fact, having been found since the early fifties, and it is therefore only reviving as the new decentralization law implemented in early 2001. As the implementation of the 1999 laws perceived by the central government as creating the problem of conflicting authorities, the Megawati government began to initiate a revision to the 1999 autonomy laws, and in 2004, a revised autonomy law (No. 32, 2004) was introduced. The controversies and different interpretation to the 1999 regional autonomy laws reflect the continuing contestations in striking a political balance between the proponents of strong centralization on the one hand and the advocate of maximum decentralization on the other hand. Although the process of moving away from a centralistic government model of Suharto's New Order would likely be irreversible, the main question remains as to where be it heading and what is changing?

New Politics?

The new politics in the post-Suharto era that is labeled by its advocates as the *reformasi* or reform politics apparently characterize two important features. The first character is its strong desire to radically reformat the political system primarily through the amendments of the constitution. It is the first time since independence that political momentums were gathered and provide an avenue to tackle the root causes of the New Order's political impasse by changing radically the political system. Several changes were undertaken, such as restructuring the position of the major state's political institutions: parliament, the presidency, and general election. The second character is manifested in a powerful drive to geographically distribute the state power into the regions. A new decentralization politics is therefore reborn after it was begun to be frozen at the end of 1950s with the success of the military onslaught to the regional rebellion in Sumatra and Sulawesi (PRRI/Permesta). The political reform is surely bringing new hope to its capacity for the reduction in horizontal inequalities.

Inequality is strongly connected with the way the accumulation of power and resources has been heavily concentrated in the capital and core regions, especially Java. Since the ending of the regional challenge in the late 1950s, power and resources have been flown from the Outer Islands to Java. A process of sucking up of rich natural resources mostly located outside Java is occurring that resulted in the increasing economic inequality between Java and the other islands. The increasing economic inequality is also followed by the process of brain drain as employment and education are mainly growing in Java rather than in the Outer Island. While the educated population is moving to Java, the poor and unskilled population were transmigrated by the government to the Outer Island that makes the Outer Island becoming the dumping area for the unwanted population. In the pocket of rich natural resources in the Outer Island, however, people from different places—most notably people belong to ethnic groups that are well known as having a strong propensity to move (Bugis, Madurese, Banjar, Minangkabau, Torajans, and so on) are flocking to seek their fortune, and become strong competitors for the local population. In this situation, geography (Java vs. other islands) and demography (changing in ethnodemographic composition) is the underlying basic structure of the society that will eventually influence the development of a new pattern of cultural identification group configuration that shows substantial differences in its social, economic, and political achievements between different regions. The crucial issue that should be investigated in this new sociocultural pattern is whether in the culturally defined group—most notably ethnicity and religion—is overlapping with their social, economic, and political achievement in their respected geographical territories.

While various sources have confirmed the increasing vertical inequality that widening the gap between the different socioeconomic classes across the board, its possible trajectory with the horizontal inequality as indicated by the process of social, economic, and political marginalization of the local populations would likely be instigated horizontal tension between different culturally defined groups. Spatial mobility between different islands and geographic territories has been instigated by the state's policies and the economic markets. The likely crossing path and trajectory between vertical and horizontal inequalities are evidently shown by the violent conflicts in Sambas West Kalimantan between the Madurese migrant community and the local Dayak populations in December 1996 until February 1997. This communal conflict occurred before an economic crisis is started, and Suharto still in control of political power. While some observers resort their arguments into the conventional cultural explanation of the violent prone behavior of both the Madurese and the Dayak, it also becomes more apparent that socioeconomic inequalities have generally widening up between regions, economic classes, and culturally different groups as many studies since the beginning of the 1990s have shown. Ruth McVey succinctly commented in 1996 on the situation in Indonesia as the following: "By the mid-1990s the combination of intense emphasis on social stability, state control, and capitalist development has resulted in what might best be described as dynamic paralyzes."[6] The paralyzes indeed finally end Suharto from power; however, their political ramification goes further beyond Suharto.

The increasing combination pressures of post-Suharto's politics and continuing economic crises that make the processes of constitution's amendment and the formulation of new decentralization policy were rapidly moving. The drafting of two decentralization laws in 1999 and their implementation in 2001 substantially reformat the national political structure in which the local elites could play more important roles. It is precisely this perceived promised from the new regional autonomy laws that triggered a long waited local enthusiasms that further express in various forms of political articulation. Among the local expression is the widespread demand to speed up the process of creating new province and district particularly as the local elites perceive the economic and political opportunities behind this regional reclassification under the umbrella of regional autonomy. As the statistics have shown, after the enactment of the regional autonomy laws, the number of provinces and districts is increased. The number of provinces increased from 27 to 33, and the number of districts drastically increased from 292 before 2001 to 434 in 2003.

The motivation to establish a separate province or district could have been initiated long before the introduction of the new autonomy laws. Under the new decentralization laws, however, local leaders or communities were aspiring to have their own administrative territories becoming more intrigued to realize their goals. While a process of political lobbying to get approval from the local parliament to endorse the demand for creating new province or districts is the first important step, it must be followed by process of getting agreement from the parliament at the national level in Jakarta before the proposal is finally signed by the president. The process of political negotiation among the elites, at the local and the national levels, is a political process that occurs beyond the everyday politics at the grassroot level. The grassroots, however, is a crucial stakeholder as their aspirations are needed by the political elites in justifying their proposal to create a new province or districts. In this context of local elite's effort to obtain the grassroot supports that political

---

[6]See Ruth McVey. 1996. "Building Behemoth: Indonesian Construction of the Nation-State," in Daniel S. Lev and Ruth McVey (eds,) *Making Indonesia: Essays on Modern Indonesia in Honor of George Mc T. Kahin.* Ithaca: Cornell Southeast Asia Program.

mobilization is very crucial as cultural identity and locality can easily be used and manipulated.

The incidence of communal conflicts in Kalimantan (Sambas and Sampit), Sulawesi (Poso and Tentena), and Maluku (Ambon and Halmahera) involving different ethnicities and religious group identities posits a possible link between the political elite's effort to create separate administrative territories and conflicts. In other places where the process of creating a new province or district proceeds without violent conflict, cultural identities apparently have been played out in creating the difference between "we" and "them." The process separating Banten from West Java (Bantenese vs. Sundanese) and Gorontalo from North Sulawesi (Christian Minahasan vs. Muslim Gorontalo) is a clear example that cultural identity that overlaps with locality is proven to be strong justifications for the local elites in establishing new provinces. Currently, a political negotiation is taking place among political elites in East Kalimantan to create North Kalimantan Province and in Central Sulawesi to create East Sulawesi Province. North Kalimantan Province will consist of four districts (Tarakan, Bulungan, Malinau, and Nunukan); and Sultim (Sulawesi Timur), while East Sulawesi Province will consist of five districts (Poso, Morowali, Tojo Una-Una, Banggai, and Luwuk). According to the revised regional autonomy law (Law No. 32, 2004), to create a new province at least five districts are required. Previously, under Law No. 25, 1999, it required only four districts to form a new province. At present, the local elites in East Kalimantan that want to create North Kalimantan Province, is facing a problem as they currently only have four districts that agreed to split from East Kalimantan Province. In justifying their goal, they argue that the central government should consider their geostrategic location in the borderland with Malaysia, where potential external threats to the national integration are at stake. The diplomatic failure to prevent two small islands (Sipadan and Ligitan) in 2002 from Malaysian claim and the current dispute on the location of oil exploration in Ambalat with Malaysia as a case in point.[7]

What about the political conflicts that are originating from the demand of ethnonationalist group in Aceh and West Papua that seek for separation from the Indonesian state? The post-Suharto's government has located these two political and arms conflicts within the political format of decentralization and regional autonomy solution package. Two special autonomy laws were created in 2001 for Aceh (Law No. 18) and Papua (Law No. 21) that recognize special status for Aceh and Papua; it is also intended to reduce the economic and political inequalities that are experienced and perceived by the people in Papua and Aceh. The creation of special autonomy laws to solve the local grievances in the province of Aceh and Papua basically contains a similar recipe in which territorial reclassification has become the core central government's solution based on the argument that the public services are too

[7]As a province, these local elites argued that it is entitled to have their own Kodam (Komando Daerah Militer or Regional Military Command). See the report in *Kompas*, June 8, 2005, *Membentuk Provinsi Kalimantan Utara demi Menjaga Kedaulatan* (creating North Kalimantan Province for Guarding the Sovereignty).

limited to cover a large administrative territory. Apart from the new revenue sharing, splitting Papua (and likely also for Aceh) into several provinces becomes the main proposal within the special autonomy laws. In the case of Aceh, an Islamic Laws are becoming an important component within the special autonomy package given by the central government. While some observers expressing their concerns on the hidden political agenda of the central government behind the provision of special autonomy status for Aceh and Papua, it has still to be seen, however, how the implementation of these laws will be able to actually reduce the real and perceived inequalities between different groups and between the center and the region.[8]

In the analysis of political development, demography—in particular, its composition by ethnicity and geographic distribution—is rarely considered as having an important influence in the dynamics of politics both at the national and local levels. The recent political development in which decentralization has become the most contested notion Indonesian observers also failed to understand how the politics of decentralization and the violent communal conflicts that followed are in fact cannot be understood without seriously considering the politics of demography that is deeply involved. Every region—province or district—is a small part of a noticeably big country, namely Indonesia. Although it seems to become truism that the communal conflict that occurred—religious or ethnic—is "highly locally concentrated."[9] Isolating communal conflict and perceive it mainly constituting local dynamics however would be misleading as the event clearly show the interconnectedness between the national decentralization politics and the local responses. The implementation of new regional autonomy laws has restructured the distribution of power and authority in the region. These new political situations open the opportunity for the local elites to manipulate locals' grievances to achieve their economic and political short-term goals. The expression of local grievances is varied according to their differences in the dynamics of its local politics that, in some instances, influence by the ethnodemographic configuration of the society and population.

---

[8] Apart from "horizontal" communal conflict in Kalimantan, Sulawesi, and Maluku, several bombs blasted in Bali (2002) and Jakarta (2003), in which a group of radical Muslim—the Jamaah Islamiyah (JI)—loosely connected with Al Qaeda have been implicated as the terrorist group behind these attacks (see ICG reports on JI network in Indonesia and Southeast Asia). Sydney Jones, the ICG director for Indonesia, indicated that JI connections were operating behind the bombing in Tentena. With regard to the separatist ethnonationalist movement, in August 1999, East Timor Province become an independent country, while secessionist movements in Papua and Aceh also have gathered momentum to press Jakarta to pay more attention to their political aspirations. In the case of Aceh, the tsunami has apparently created a new political climate that brought together Indonesian government representatives and GAM to sit together for a series of mutual dialog sponsored by the former Finish prime minister and now the head of Crisis management initiative (CMI) Martti Ahtisaari. The fifth and last round of the meeting was conducted in August 2005, in which a formal draft agreement was signed by the two parties.

[9] Ashutosh Varshney, Rizal Panggabean, and Mohammad Zulfan Tadjoeddin, 2004, "Patterns of Collective Violence in Indonesia (1990–2003)." UNSFIR Working Paper-04/03.

The Politics of Change Without Change in Politics

Politics in Indonesia after Suharto have gone into the seemingly new politics in which the member of the society and the regions can articulate their political aspiration with greater freedom than before. The state management no longer in the hand of one person or the central government but has been dispersed into a diverse political organization, most notably the parliaments and the local government. A series of new amendments to the constitutions have been ratified that provide a new basis for the creation of a new political system that, in theory, should have a better chance to accommodate the people's needs rather than only serving the ruling elites. The new political system that is characterized by a strong drive for decentralization has proliferated local enthusiasms as reflected in the creation of the substantial number of new districts. The creation of new territorial boundaries as a response to the new regional autonomy laws, however, could be seen as a process of the fragmentation of the nation-state. In this process of internal disintegration, an ethnodemographic configuration is becoming a crucial factor that is potential for political manipulation.

At present, local enthusiasm also is shown with the implementation of direct election of the regional government's head—both at the province and district levels. Again, as clearly indicated by the evidence in several districts, the pattern of political mobilization is interestingly still very much dependent on the old political pattern in which the local elites become the primary agent, and the people are merely political onlookers. The new political format that is expected to be the political platform for creating a social transformation that strengthened people's interest apparently not materialized as the politics still largely control by the local elites that mostly representing the old political players. The horizontal inequality that is expected to be reduced by the implementation of a new political system and the expansion of regional autonomy generally not materialized. Reassertion of ethno-communal rights that is part of the local responses to the new political opportunities easily manipulated by the local elites in their strategy to grab the political position and the short-term economic benefits.[10] As evidently shown in several districts, for example, in Sambas, West Kalimantan, and Kotawaringin Timur, Central Kalimantan, and Poso in Central Sulawesi, communal conflicts between different culturally defined groups are closely linked with the local elites contestation for a local government position that occurs in the context of district territorial reclassifications known as *pemekaran* (literally means "blossoming" or "blooming") and the creation of new districts.

Indonesia that is born with a strong spirit of civic nationalism that promises to enhance plurality and differences is currently facing new challenges to survive its ideal. Primordial sentiments that began at the beginning of the twentieth century as

[10]A recent study describes it as bureaucratic rent-seeking behavior, and see Fitria Fitrani, Bert Hofman, and Kai Kaiser, "Unity in Diversity? The creation of New Local Governments in a Decentralising Indonesia," *Bulletin of Indonesian Economic Studies,* Vol. 41, No. 1, 2005: 57–79.

the new impetus for the educated indigenous population to form social and political organizations seem to revive in the beginning of twenty-first century in the guise of decentralization and regional autonomy. The emergence of primordial sentiments, particularly in the form of ethnicity, was largely expressed through the intention to have separate administrative territory, either province or district, that resembles their perceived cultural identities. In these political circumstances, violent conflicts in West Kalimantan, Central Kalimantan, Central Sulawesi, and Maluku, as well as in Aceh and Papua, have been perceived by the political elites in the region and in the center to justify their political intention to create new territorial–administrative boundaries.

The political elite motivation to create a new province or district is varied and, in many instances, reflects a mixture of motives from purely personal ambition for having a political and bureaucratic position to the aspiration of developing a modern society in the region. In realizing their intention, ethnicity has provided an avenue that could be manipulated to mobilize people's support and, at the same time, a cultural legitimating medium to justify their motives. In these political trajectories, ethnicity has been juxtaposed for the short-term goals of the local politics and economic elites. The expression of ethnicity through the sentiment of being attached to locality or being a member of cultural group identity or being part of local history or the combination of those factors could have been perceived as an instrument ready for political orchestration. It is indeed such ethnic orchestrations that lead the current politics of decentralization into another Indonesia's political abyss. The expectation that political reform will result in the political changes and democratization apparently not materialized instead, we are witnessing a spiraling of political events that might be best described as the process of "involution."[11] Borrowing from Clifford Geertz's famous "agriculture involution" thesis to explain the process of increasing complexities on the structure of political practices in Indonesia without making any significant changes on its substance. In a similar vein, Ignas Kleden—the leading Indonesian thinker—describes the present situation in Indonesia as "the politics of change without a change in politics."[12]

Concluding Remarks

Two internal processes are currently threatening Indonesia's survival as an integrated nation-state. The first is the process of national mystification of the *NKRI* that works from "above," and the second is the process of ethnoterritorial slicing that is working from "below." Such a slicing process will likely turn Indonesia into a fragmented state and failed in enhancing plurality and equality. The internal threats

[11] See Clifford Geertz's. 1963. "Agriculture Involution." *The Processes of Ecological Change in Indonesia.* Berkeley: University of California Press.

[12] See Ignas Kleden's column, *Tempo,* February 13, 2005, pp. 56–57.

that are currently working from within are generally overlooked by the ruling elites. The political elites are preoccupied with the perception that the threats to national integration are coming from outside as the work of Indonesia's external enemies. The usual blame on the external threats and the failure to realize the actual threats that are created internally will escalate the process of disintegration from within. The present politics of decentralization has been hijacked by the political elites to serve their short-term economic and political goals.

The mystification of the "final-conclusive unitary state" is particularly derived from the dominant but narrow military thinking of national integration. The present repressive military approach in solving the problem in Aceh and Papua also essentially reflects the process of disintegration from within. Sooner or later, internationalization is unavoidable political imperatives in the process of resolving the conflict in Aceh and Papua. In the current fast-moving communication era, any state will no longer have the luxury to close its borders and to ensure its conventional sovereignty. Global interconnectedness will trespass the nation-state's porous boundaries. Indonesia is no exception. Reinventing civic nationalism that upholds differences is perhaps what the Indonesian people should do to save their future from the mystification of its unitary state formation and from the "enclavization" of its decentralization politics.

Epilog

Indonesia's Political Demography Challenges in the Century

In an op-ed essay for *The Jakarta Post* on 25 June 2015 I wrote the following:

> Demography will remain one of toughest challenges for Indonesia's future. The silent nature of demographic pressures makes the general public unaware of its profound impacts, economically and politically. Managing demographic pressures requires a proper understanding of demographic characteristics and their interplay with other variables.[1]

Now, 7 years after that essay which aimed to alert President Jokowi on the challenge brought about by demography, the circumstances seem unchanged, even getting more critical. As the fourth largest country in the size of the population after China (1.44 billion), India (1.37 billion), and USA (330 million), Indonesia (270 million) faces even a greater tremendous challenge in managing the sustainability of its social, economic, and political developments in such a far-flung territory.

Proclaimed as a sovereign state in 1945 following the end of World War II, Indonesia joined the newly decolonized states in Asia and Africa, such as India, Pakistan, Burma, Kenya, Egypt, and Ghana. As world history moving toward the bipolar global politics of the Cold War, Indonesia's first president, Soekarno, initiated a political campaign to create a new political block of non-aligned countries mostly comprising the newly decolonized nation in which China and Vietnam were also part of it. The creation of a non-aligned movement initiated by Soekarno shows the strong aspiration of Indonesian leaders to contribute to shaping the global political order that is often dictated by superpowers like the USA and at the time the Soviet Union.

As a country that for around 300 years lived under the Dutch colonial power, Indonesia has learned a long history of subjugation and struggle to be an independent nation. In this important historical context, the beginning of the twentieth century

[1] "Managing population under Jokowi," *The Jakarta Post*, 25 June 2015.

R. Tirtosudarmo, *From Colonization to Nation-State*,
https://doi.org/10.1007/978-981-16-6437-3

could be seen as the significant era in which a seed of national consciousness began to flourish, epitomized by Kartini. Kartini was born in 1879 as a Javanese daughter of a regent in Pati, Central Java. Kartini, through letters written in the Dutch language to her pen friends in Holland, marks the beginning of a sense of being native that its life controlled by others more powerful. The letters also represent the voice of a girl that is treated differently from her male siblings by its own Javanese tradition. Her letters that are then published in Holland symbolize the cry for freedom both as the woman and the native people.

One thing is certain that what has been developing for a century is beyond what the Dutch colonial government imagined. The ethical policy that the goal is to uplift the welfare of the native proves to be affecting not only the social and economic conditions of the native but more importantly have changed the political aspiration of the view educated native people. Through higher education that originally aimed to train the native in several vocations needed to support the social and economic programs, such as public health, engineering, and legal bureaucracy; in short, to produce the professional groups that could strengthen the technocracy of the colonial government. What is unexpected is the impact of the penetration of the capitalist system through the various introductions of modern economic institutions and infrastructures, among others through the introduction of the labor union and the print media. As Benedict O.Gorman Anderson has meticulously written in his famous book *Imagined Communities: On the Origin and the Spread of Nationalism*, the impact of print capitalism is profound. He shows how the shaping up of nation-state attributed to more by national consciousness triggered by colonial policies and civic engagement, rather than originated from being the member of ethnic communities.[2]

A sense of belonging to a wider community or nation began to be felt in the dawn of the twentieth century. As argued in the first chapter in this book, the end of the nineteenth century and the beginning of the twentieth century were marked by several important social changes, the impact of which significantly altered the course of history in Indonesia. The implementation of an ethical policy aimed to improve the welfare of the native through education, irrigation, and emigration has become the pinnacle of a major transformation in the life of the native. In this book, emigration or colonization manifested in the policy is to balance the distribution of population between Java that is perceived as already overpopulated and other islands that are perceived as empty continued after independence. Transmigration in a nutshell exemplified the critical idea of political demography in the nation-state building processes. The colonization of islands outside Java through the creation of Javanese agricultural enclaves also reflects the obsession among the ruling elites after the independence of demographic engineering as a device to achieve national integration. The book that consists of 12 chapters offers a discourse as well as

[2]Benedict O. Gorman Anderson. (1983). *Imagined communities: The origin and spread of nationalism*. London: verso.

empirical evidence of how a nation-state is in the making or unmaking in the last 100 years.

What should also be given attention to is the expansion of urban areas as economic activities continue flocking in the cities, especially in Java. The number of the younger generation increased rapidly, which triggered the high demand for higher education. Again, as the good higher learning facilities are mainly concentrated in Java, it caused more and more migration of young people toward Java, making Java the center of education. Cities in Indonesia as well as in many other countries have become the showcase in which the movement of people across the globe is contested. In this context, a recent book entitled *The City in Transgression: Human Mobility and Resistance in the 21st Century* locates human mobility in its crucial perspective:

> As soon as I had some sort of understanding of the complex issues concerning global human mobility, new sets of circumstances and conditions would come to the fore. The endless string of human catastrophes that highlight the lengths to which refugees, asylum seekers, and migrants would go—lives lost in deserts, at sea, in transit in refugee camps, in detention camps, suffocating in refrigeration trucks, enslaved, bought, sold, and sexually abused in conflict zones. These are just some of the harrowing experiences amid this unfolding human tragedy.[3]

The main thesis of the book is the shaping up of the nation-state from the perspective of political demography in which movement of people engineered or voluntary is perceived as a critical variable in such process. The archipelagic nature of the Indonesian geographical landscape in which Java, the most populated island, constitutes the center of gravity attracts people to migrate for economic and other reasons. As Indonesia embarks on the new political regime after 1965 that emphasized economic development, the government began to control the growth of the population as well as its geographic distribution. Two main national population policies are created, a family planning program to reduce the birth rate and a transmigration program to relocate people from densely populated islands, namely Java and Bali, to Sumatra, Kalimantan, Sulawesi, and Papua. The graph below (Fig. 1a) shows that since 1971, the rate of population growth successfully reduced from 2.1 to 1.25 in 2020. Yet, the Indonesian population continues to growth from 11.21 million in 1971 to 270.20 million in 2020.

As the birth rate significantly falls in the past 50 years, the age structure of the Indonesian population also changes considerably. The following graph (Fig. 2) indicates the changing age structure since 1971, showing the declining proportion of population age 0–14 and the increased proportion of population age 15–64 and 65+. The demographic situation reflects what is in the literature popularly known as the periods of demographic bonus in which the proportion of the population in the working age (15–64) constitutes the largest group. The challenge for the policymakers is therefore whether sufficient jobs could be created to accommodate

[3] Benedict Anderson. (2020). *The city of transgression: Human mobility and resistance in the 21st Century*. London: Rutledge.

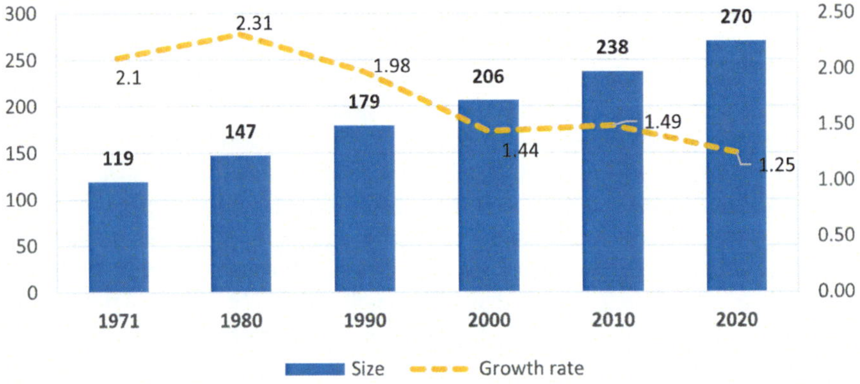

Fig. 1 Indonesia: the size and growth rate of population, 1971–2120. (Source: Muhidin 2021)

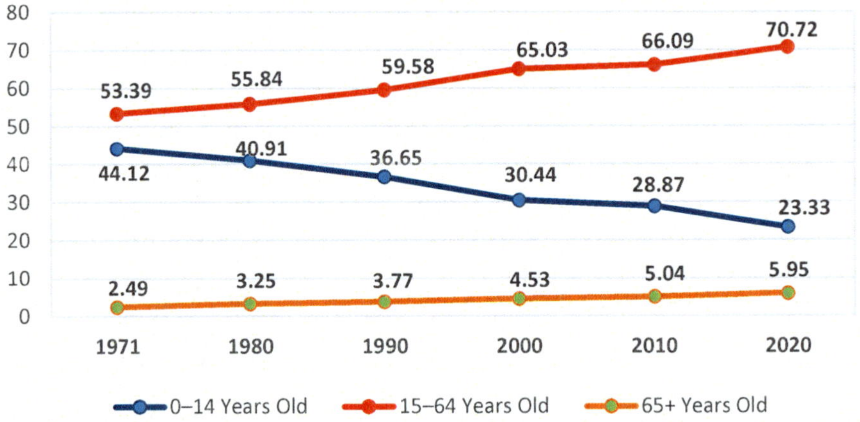

Fig. 2 Indonesia: the age structure, 1970–2020. (Source: Muhidin 2021)

the increasing demand, especially from the younger generation that moves from education institutions to the marketplace. The social and political ramifications of the unmet demand for the working age population, especially the youth, could be shattering as the unemployed youth may easily be attracted by political entrepreneurs to engage in uncivil activities, such as radical and terrorist groups. Since Covid-19 began to spread in February 2020 as in many other countries, the Indonesian economy has also been deeply affected. After more than a year, while vaccination provides hopes to halt the spread of the pandemic, the slowdown in the economy adds to the problem of creating new jobs for the increasing working age population.

Looking back, if 1971 is considered as the beginning of the settlement of the New Order regime describe in 1978 by Herberth Feith, the Australian scholar, as the repressive developmental regime, Indonesia has gone through its 50 years long complex development.[4] Yet, we saw the New Order regime under President Suharto finally ended in 1998 following what American scholar, Ruth McVey, described as its internal paralysis.[5] McVey succinctly writes the predicament of the New Order of Suharto as the following:

> By the mid-1990s the combination of intense emphasis on social stability, state control, and capitalist development has resulted in what might best be described as dynamic paralysis.

However, the political development after the demise of Suharto was not as people expected. The euphoria of what is popularly known as reformation proves to be mere words, only rhetoric as the political structure continues unabated. Regime changes, but not the structure of the body politics. In the words of Robison and Hadiz (2004), what happened is just a reorganizing power.[6]

Most of the population and the masses continue unable to improve their economic condition as promised by political elites that heralded the reformation. The fact is people move mostly in a short distance rather than long distance, let alone crossing the border to other countries. Therefore, different from my previous book that dealt with international migration, this book is more about internal migration, migration within Indonesia that mostly engaged in a short rather than long-distance movement.[7]

Available data on the distribution of the population by provinces across the country despite the decreasing number of the population residing in Java yet as Table 1 shows only a reduction of less than 5% between 1980 (61.88%) and 2020 (56.10%). The 5% decline of the population in Java between 1980 and 2020 distributed into other islands, namely Sumatra (19–21.67%), Nusa Tenggara (5.38–5.54%), Kalimantan (4.56–6.15%), Sulawesi (7.06–7.36%), and the rest distributed in places such as Maluku and Papua. However small it looks from the perspective of the national level, it could be remarkably high if we look at the perspective at the local level, for example in the case of Papua. Migration to Papua is one of the hot issues as it creates a problem of competition between migrants and the local Papuans in search of dire economic opportunities, particularly in urban areas. Table 1 shows that Indonesia is increasingly becoming more urbanized between 1980 and 2020 as more and more people reside in the cities. Again, there is a regional difference on the increase of urbanization in which Java shows the highest (62.6%) followed by Sumatra (41.4%), Nusa Tenggara (42.6%), Kalimantan

---

[4]Herbeth Feith. (1980). Repressive-developmentalist regime in Asia: Old strengths, new vulnerabilities. *Prisma,* No. 19, December, pp. 39–55.

[5]Ruth McVey. (1996). Building Behemoth: Indonesian construction of the nation-state. In Daniel S. Lev and Ruth McVey (eds.), *Making Indonesia: Essays in modern Indonesian honor of George Mc T. Kahin.* Ithaca, New York: Cornell Southeast Asia Program.

[6]Richard Robison and Vedi R. Hadiz. (2004). *Reorganising power in Indonesia: The politics of oligarchy in an age of market.* London and New York: Routledge Curzon.

[7]See Riwanto Tirtosudarmo. (2018). *The politics of migration in Indonesia and beyond.* Singapore: Springer and LIPI Press.

Table 1 Indonesia: population distribution by province, 1980–2000

Island/Province	Percentage distribution of population					Density	Urban	Median	Ratio	Growth %
	1980	1990	2000	2010	2020	(per KM2)	(%)	Age dependecy 2010–2015		
						(Year of Reference 2015)				
Indonesia	147,490	179,379	206,265	237,641	270,204	134	53.3	28.5	49.20	1.43
%	100%	100%	100%	100%	100%					
Sumatra	19.00	20.35	21.00	21.31	21.67	105	41.4	23.6		1.73
Aceh	1.77	1.90	1.91	1.89	1.95	86	30.5	24.7	54.89	2.13
North Sumatra	5.67	5.72	5.65	5.46	5.48	191	52.6	24.8	56.90	1.41
West Sumatra	2.31	2.23	2.06	2.04	2.05	124	44.2	23.2	55.84	1.38
Riau	1.47	1.84	2.40	2.33	2.37	73	39.6	24.6	52.46	2.71
Riau Island				0.71	0.76	240	83.0	23.0	48.45	3.23
Jambi	0.98	1.13	1.17	1.30	1.31	68	32.0	22.7	47.30	1.90
South Sumatra	3.14	3.52	3.35	3.13	3.13	88	36.5	23.0	49.62	1.54
Bangka-Belitung			0.44	0.51	0.54	83	52.5	22.2	47.25	2.30
Bengkulu	0.52	0.66	0.76	0.72	0.74	94	31.7	22.9	47.99	1.76
Lampung	3.14	3.35	3.27	3.20	3.33	234	28.3	22.0	49.67	1.28
Java	61.88	59.97	58.83	57.49	56.10	1055	62.6	29.9		1.19
Jakarta	4.41	4.60	4.07	4.04	3.91	15,292	100.0	20.6	39.41	1.11
West Java	18.61	19.73	17.32	18.12	17.87	1319	72.9	21.8	48.24	1.63
Banten			3.93	4.47	4.41	1235	67.7	23.2	46.94	2.34
Central Java	17.20	15.90	15.14	13.63	13.51	1029	48.4	28.4	48.61	0.83
Yogyakarta	1.87	1.62	1.51	1.45	1.36	1173	70.5	27.0	44.52	1.23
East Java	19.79	18.12	16.86	15.77	15.05	812	51.1	27.5	44.44	0.71
Nusa Tenggara	5.38	5.25	5.39	5.50	5.54	179	42.6	23.2		1.50
Bali	1.67	1.55	1.53	1.64	1.60	718	65.5	28.5	45.47	1.29
West Nusa Tenggara	1.85	1.88	1.94	1.89	1.97	260	45.4	23.5	54.91	1.42
East Nusa Tenggara	1.86	1.82	1.92	1.97	1.97	105	21.6	16.7	69.30	1.77

Kalimantan	4.56	5.07	5.49	5.80	6.15	25	44.8	23.0		2.11
West Kalimantan	1.69	1.80	1.96	1.85	2.00	32	33.1	23.5	52.67	1.70
Central Kalimantan	0.65	0.78	0.90	0.93	0.99	16	36.6	23.1	48.17	2.40
South Kalimantan	1.40	1.45	1.45	1.53	1.51	103	45.1	22.3	49.57	1.90
East Kalimantan	0.83	1.05	1.19	1.27	1.39	27	66.0	22.9	46.34	2.48
North Kalimantan				0.22	0.26	8		23.9	52.15	4.04
Sulawesi	7.06	6.98	7.25	7.31	7.36	92	36.9	23.2		1.48
North Sulawesi	1.43	1.38	0.98	0.96	0.97	174	49.8	20.1	46.66	1.20
Gorontalo			0.40	0.44	0.43	101	39.0	23.8	50.11	1.70
Central Sulawesi	0.87	0.95	1.08	1.11	1.10	46	27.2	22.7	52.02	1.74
South Sulawesi	4.11	3.89	3.91	3.38	3.36	182	40.6	22.9	54.06	1.16
West Sulawesi				0.49	0.53	76	22.9	15.2	59.58	2.01
Southeast Sulawesi	0.64	0.75	0.88	0.94	0.97	66	31.2	16.0	60.43	2.25
Maluku	0.96	1.04	0.58	0.65	0.68	36	38.0	16.0	60.82	1.89
North Maluku			0.38	0.44	0.47	36	27.8	15.8	60.95	2.25
Papua	0.80	0.92	1.08	1.19	1.59	10	28.4	15.4	52.08	2.10
West Papua				0.32	0.42	9	32.3	15.0	52.08	2.70

Source: Muhidin (2021)

(44.8%), Sulawesi (36.9%), Maluku (36%) and the lowest are found in Maluku Utara (27.8%), Papua (28.4%), and Papua Barat (32.3%).

What is beyond expectation is the sudden spread of coronavirus that was firstly found in Wuhan, China. The deadly virus called Covid-19 in a short time has become a pandemic that sweeps into all countries without exception. The spread of the virus through people movement forces all the governments to control the movement of people both within and between countries. Demography, migration particularly, immediately becomes the human behavior that needed to be controlled to prevent the spread of the virus. It has been more than a year since the first spread of the virus from Wuhan. While the vaccine to prevent the disease has been found, the vaccination for the endangered population took time to really contain the pandemic.

In the history of the world and the history of Indonesia as well, pandemic is not uncommon. It has happened during colonial times as the pest rampage in a place like Malang in East Java. Yet the Covid-19 pandemic seems unparalleled as it occurred at such a speed as migration becomes the main feature of twenty-first century globalization. While globalization in the early twentieth century occurred in the context of western colonization, in the early twenty-first century, globalization occurred simultaneously with the spread of market capitalism through the expansion of transportation and communication technology. As the fourth largest country in the world in the size of its population, Indonesia is confronted by a real challenge to secure the population from the spread of this deadly virus.

Now, in the beginning of the third decade of the twenty-first century, the question is what has been learned and not learned; and what likely are the current and future challenges for Indonesia's political demography to continue its development course as a nation-state. There seem to be some parallels as well as paradoxes when we place Indonesia in perspective of what takes place in the early twentieth and twenty-first centuries. Among the parallels: (1) The state, both the colonial and the post-colonial, continues to be confronted by the problem of demographic imbalance between "inner" and "outer" islands, between Java and other islands; (2) in the early twentieth century foreign capital began to take place in the economy, apparently similar to what is emphasized under the current government; (3) the state, both colonial and the current government, shows similar characteristic as authoritarian and undemocratic with the obsession to develop modern public infrastructures; (4) increasing problem of economic inequality and social injustice between the limited few and the majority of the population; and (5) the challenge of maintaining the territorial integrity as centrifugal tendencies continually recurred, at the beginning of the twentieth century the Dutch struggle to conquer Aceh, and now the Indonesian government confronting separatist movement in Papua.

History, as many pundits have said, is repeated itself. Indonesia, as the book exposed, could be a showcase of how population dynamics and political change intertwined. And as indicated seven years ago, "managing demographic pressures requires a proper understanding of demographic characteristics and their interplay with other variables." In this interplay one critical demographic variable that as strongly argued in this book as to have a significant role in shaping the nation-state of

Indonesia is movement of people. Colonization as the colonial state's demographic engineering policy to move people from Java to "outer islands" continued by transmigration policy after independence. As the book intensively discussed the continuing policy provide the way in shaping up the different communities into an integrated nation. Yet as the book has shown the final result of the nation-state building in this far-flung archipelago continues to be contested and as persuasively indicated above might be more convoluted in the twenty-first century.

Source of Chapters

Part I

Chapter 1. Originally a working Paper Series, No. 97–05, 1997, Population Studies and Training Center, Department of Sociology, Brown University, USA.

Chapter 2. Originally published as a working paper at National Center for Development Studies (NCDS), Australian National University (ANU), 1990.

Chapter 3. A revised version of two articles published in Prisma (English edition), December 1989 and in Masyarakat Indonesia, No. 1, 1991.

Part II

Chapter 4. A revised version of an article published in *Masyarakat Indonesia*, No. 3, 1985.

Chapter 5. A revised version of a paper published in Population Movement in Southeast Asia: Changing Identities and Strategies for Survival, edited by Kenichi-Abe and Masako Ishii, pp. 177–194. The Japan Center for Area Studies (JCAS) Series 10, National Museum of Ethnology, Osaka, 2000.

Chapter 6. Originally published at Indonesian Anthropology Journal, Volume XXVI, No. 67, pp. 30–45, January–April 2002.

Part III

Chapter 7. Originally published as a chapter in Labour Migration in Indonesia: Policies and Practice, edited by Sukamdi et al. pp. 91–116, Yogyakarta: Population Studies Center Gajah Mada University.

R. Tirtosudarmo, *From Colonization to Nation-State*,
https://doi.org/10.1007/978-981-16-6437-3

Chapter 8. A revised version of a paper presented at the workshop on Labor Migration and Socio-Economic Change in Southeast and East Asia, 14–16 May 2001, Lund, Sweden.

Chapter 9. Paper originally prepared for a conference on "Crossing Borders, Traversing Boundaries: Bridging the Gap between International and Internal Migration Research and Theory", 13–14 October 2011, Asia Research Institute, National University of Singapore.

Part IV

Chapter 10. A revised version of a paper presented at Annual Lecture Series, Center for Southeast Asian Studies, Monash University, Melbourne, Australia, June 8, 2002.

Chapter 11. This paper is originally prepared for the 2013 SIEAS Research Cluster Conference, *Southeast Asia vis a vis Estrangers in the Historical Perspective,* Sogang University, Seoul, 31 October–1 November 2013. A slightly revised version appears as an article in *Jurnal Kajian wilayah (Journal for Area Studies),* Vol. 6, No. 2, 2015, pp. 161–172.

Chapter 12. A revised paper presented at CRISE workshop, Oxford University, 29–30 June 2005.

Bibliography

Abdullah, Firdaus Haji. 1993a. The Rumpun Concept in Malaysia-Indonesia Relations. *The Indonesian Quarterly* XXI (2): 137–150.

———. 1993b. The Phenomenon of Illegal Immigrations. *The Indonesian Quarterly* XXI (2): 171–186.

Abdullah, Moha Asri. 1997. The Inflow of Foreign Labor to Malaysia: Some Critical Analysis of Socio- Economic and Political Implications on the Locals. Paper presented at the 12th New Zealand International Conference on Asian Studies, Massey University, Palmerstone North, 26–29 November.

Acciaioli, Gregory L. 1998. Bugis Entrepreneurialism and Resource Use: Structure and Practice. *Antropologi Indonesia* 57: 81–89.

Achmad, I. Amal. 1984. West Sumatra and South Sulawesi: Two Regions in Their Relation to Central Government, in Indonesia. Unpublished Ph.D. thesis, Monash University.

Adam, Asvi Warman. 1994. Pengiriman Buruh Migran Jawa ke Vietnam tahun 1900-an (Exporting Javanese Migrant Workers to Vietnam in the 1900s). *Sejarah* 5: 1–6.

Adi, Rianto. 1995. *Migrasi Internasional Tenaga Kerja Indonesia: Harapan dan Kenyataan.* Jakarta: Pusat Penelitian Universitas Katolik Atma Jaya.

———. 1996. The Impact of International Labor Migration in Indonesia. Unpublished Ph.D. thesis in Population and Human Resources, Department of Geography, Faculty of Arts, the University of Adelaide.

Aditjondro, G.J. 1986. *Datang dengan Kapal, Tidur di Pasar, Buang Air Dikali, Pulang Naik Pesawat: Telaah Dampak Migrasi Suku-Suku Bangsa Sulawesi Selatan dan Tenggara ke Irian Jaya sejak Tahun 1962*. Jayapura: Yayasan Pembangunan Masyarakat Irian Jaya (YPMD).

———. 1994. *In the Shadow of Mount Ramelau: The Impact of the Occupation of East Timor.* Leiden: Indonesian Documentation and Information Centre (INDOC).

Agustiar, Memet. 1998. Migrasi Penduduk Kalimantan Barat (Migration in West Kalimantan). In *Mobilitas Penduduk dan Pembangunan Daerah: Analisa SUPAS 1995*, 133–157. Jakarta: Kantor Meneg Kependudukan/BKKBN dan Lembaga Demografi UI.

Alexander, Malcolm L. 1980. Structure and Process in the Modern World-System: The World System Theory of Immanuel Wallerstein and Its Predecessors. In *An Introduction to Marxist Theories of Underdevelopment*, ed. Richard Peet, 113–122. Canberra: Department of Human Geography, Australian National University.

Alqadrie, Syarief Ibrahim. 1994. Poverty or Impoverishment? The Gap Between Apparent and Actual Reality in West Kalimantan. Paper presented at INFID Conference, 22–24 April, Paris.

Amin, Samir. 1974. Modern Migration in Western Africa. In *Modern Migration in Western Africa*, ed. Samir Amin, 65–124. London: Oxford University Press.

R. Tirtosudarmo, *From Colonization to Nation-State*,
https://doi.org/10.1007/978-981-16-6437-3

Anaf, Aswatini. 1984. Female Migration and Employment: A Case Study in Kecamatan Pasar Rebo, Jakarta. Unpublished M.A. thesis, Department of Demography, Australian National University.

Ananta, Aris, et al. 1998. The Impact of the Economic Crisis on International Migration: The Case of Indonesia. *Asian and Pacific Migration Journal* 7: 2–3.

Anderson, B.R.O'.G. 1972a. *Java in a Time of Revolution, Occupation, and Resistance 1944–1946*. Ithaca, NY: Cornell University Press.

Anderson, B.R. 1972b. *The Idea of Power in Javanese Culture*. Ithaca, NY: Cornell University Press.

Anderson, B.R.O'.G. 1985. Current Data on Indonesian Military Elite. *Indonesia* 40: 131–164.

Anderson, B.R.O.G. 2003. *Imagined Communities: Reflections on the Origin and Spread of Nationalism*. London: Verso.

Anderson, Benedict. 2020. *The City of Transgression: Human Mobility and Resistance in the 21st Century*. London: Routledge.

Arndt, H.W. 1983. Transmigration: Achievements, Problems, Prospects. *Bulletin of Indonesian Economic Studies* 19 (3): 50–73.

———. 1984. Transmigration in Indonesia. *Working Paper, No. 146*. Geneva: International Labor Organization.

Arsip Nasional Republic Indonesia (National Archive of Republic of Indonesia). 1981. *Laporan-laporan tentang Gerakan Protes di Jawa pada Abad XX (Reports on Protest Movement in Java at the 20th Century)*. Jakarta: Arsip Nasional Republik Indonesia.

Azis, Iwan J. n.d. *Impacts of Economic Reform on Rural-Urban Welfare: A General Equilibrium Framework*. Ithaca, NY: Cornell University.

Babcock, T. 1986. Transmigration: The Regional Impact of a Miracle Cure. In *Central Government and Local Development in Indonesia*, ed. C. Andrews, 157–199. Singapore: Oxford University Press.

Bachriadi, Dianto. 1998. *Merana di Tengah Kelimpahan (Misery Amidst Plenty)*. Jakarta: ELSAM.

Bachtiar, H.W. 1973. The Religion of Java: Sebuah komentar. *Majalah Ilmu-Ilmu Sastra Indonesia*, 1, Jilid V.

———. 1974. *The Indonesian Nation: Problems of Integration and Disintegration*. Singapore: Institute of Southeast Asian Studies.

Bappeda Kalimantan Selatan. 1986. Data Kalimantan Selatan Tahun 1983, 1984 dan 1985 (Data on South Kalimantan, 1983, 1984 and, Banjarmasin).

Baroto, Al. 1993. Similarities and Differences in Malaysia-Indonesia Relations: Some Perspectives. *The Indonesian Quarterly* XXI (2): 151–170.

Basri, H. 1988. Perpindahan Orang Banjar ke Surakarta: kasus Migrasi Inter Etnis di Indonesia (Migration of the Banjarese to Surakarta: A Case of Inter-Ethnic Migration in Indonesia). *Prisma 3* 18: 42–56.

Bhakti, Ikrar Nusa. 1991. The Dynamics of Indonesia-Papua New Guinea Relations: From Conflict to a Better Understanding. Unpublished Ph.D. thesis, Division of Asian and International Studies, Griffith University, Brisbane, Australia.

Biro Pusat Statistik. 2002. Kalimantan Timur Dalam Angka.

Bisri, Cik Hasan. 1985. Kalola: Sebuah Desa yang Pernah Ditinggalkan Banyak Penghuninya (Kalola: A Village Abandoned by Its Inhabitants). In *Migrasi*, ed. Mukhlis and Kathryn Robinson, 1–65. Ujung Pandang: Yayasan Ilmu-Ilmu Sosial dan Lembaga Penerbitan Universitas Hasanuddin.

Bookman, Milica Zarkovic. 1997. *The Demographic Struggle for Power: The Political Economy of Demographic Engineering in the Modern World*. London: Frank Cass.

Booth, A., and R.M. Sundrum. 1981. Income Distribution. In *The Indonesian Economy During the Suharto Era*, ed. A. Booth and P. McCawley, 181–217. Kuala Lumpur: Oxford University Press.

Breman, Jan. 1990. The Civilization of Racism: Colonial and Post-colonial Development Policies. In *Imperial Monkey Business (CASA Monographs 3)*, ed. Jan Breman, 123–152. Amsterdam: VU University Press.

———. 1997. *Menjinakkan Sang Kuli: Politik Kolonial pada awal Abad ke-20 (Taming the Coolie Beast)*. Jakarta: Pustaka Utama Grafiti dan Perwakilan KITLV.

Brunner, E.M. 1961. Urbanization and Ethnic Identity in North Sumatra. *American Anthropologist* 63: 508–521.

Budiman, Arief. 1985. *Transmigrasi di Indonesia: Ringkasan tulisan dan hasil-hasil penelitian*. Jakarta: Gramedia.

Castles, L. 1967. The Ethnic Profile of Jakarta. *Indonesia* 3: 153–204.

Castles, Stephen, and Mark J. Miller. 1993. *The Age of Migration: International Population Movement and the Modern World*. Basingstokes: Palgrave/Macmillan.

Chalmers, Ian, and Vedi R. Hadiz, eds. 1997. *The Politics of Economic Development in Indonesia: Contending Perspectives*. London: Routledge.

Chou, Cynthia. 2006. Border and Multiple Realities: The Orang Suku Laut of Riau, Indonesia. In *Centering the Margin: Agency and Narrative in Southeast Asian Borderlands*, ed. Alexander Horstmann and Reed L. Wadley, 111–134. New York and Oxford: Berghahn Books.

Choucri, N. 1984. Perspectives on population and conflict. In *Multidiciplinary perspectives on population and conflict*, ed. N. Choucri, 1–25. Syracuse: Syracuse University Press.

Christie, Kenneth, ed. 1998. *Ethnic Conflict, Tribal Politics: A Global Perspective*. Surrey: Curzon Press.

Chun, A. 1996. Fuck Chineseness: On the Ambiguities of Ethnicity as Culture as Identity. *Boundary 2* 23 (2): 111–138.

Chuzaifah, Yunianti. 1999. Krisis yang menghempas nasib buruh migrant (Crisis That Has Plunged Migrant Worker's Life). *Jurnal Analisis Sosial* 4 (2).

Coale, A.J., and E.M. Hoover. 1958. *Population Growth and Economic Development in Low Income Countries: A Case Study of India's Prospects*. Princeton: Princeton University Press.

Collier, W. L. 1979. Resource Use in the Tidal Swamps of Central Kalimantan: A Case Study of Banjarese and Javanese Rice and Coconut Producers. Paper presented at the International Symposium of Tropical Ecology, 16–20 April, Kuala Lumpur.

Collins, James T. 1999. Keragaman Bahasa di Kalimantan Barat (Language Diversity in West Kalimantan). Paper presented at Seminar Keragaman Bahasa Bumi Khatulistiwa, Pontianak, 23 September.

Cox, David. 1997. The Vulnerability of Asian Women Migrant Workers to a Lack of Protection and Violence. *Asian and Pacific Migration Journal* 6 (1): 59–76.

Cremer, George. 1988. Deployment of Indonesian Migrants in the Middle East: Present Situation and Prospects. *Bulletin of Indonesian Economic Studies* 24 (3): 73–86.

Crouch, Harold. 1986. Security Concerns Posed by Indonesia's Armed Forces. *Indonesian Issues* No. 3.

Cunningham, C.E. 1958. *The Post War Migration of the Toba Bataks to East Sumatra*, Cultural Report Series Yale. New Haven: Yale University Press.

Davidson, Jamie S. 2000. Ethnic Violence and Electoral Politics on the Periphery: The Case of West Kalimantan. Paper presented at the PER-CIK International Conference, Yogyakarta, 3–7 July.

Department of Information, Directorate of Foreign Information Services, Indonesia 1992: An Official Handbook, Jakarta (1992, pp. 55–72)

Dick, H.W. 1975. Prahu Shipping in Eastern Indonesia, Part I. *Bulletin of Indonesian Economic Studies* XI (2): 69–107.

Dick, John. 1991. *Forest Land Use, Forest Use Zonation, and Deforestation in Indonesia: A Summary and Interpretation of Existing Information*. Jakarta: EMDI Publication.

Djamin, Z. 1984. *Pembangunan Ekonomi Indonesia: Sejak Repelita Pertama (Economic Development in Indonesia Since the First Five-Year Development Plan)*. Jakarta: Lembaga Penerbit Fakultas Ekonomi Universitas Indonesia.

Djohan, Erniati. 1994. Gambaran Umum Daerah Penelitian (Overview of the Research Area). In *Pemuda, Pendidikan dan Ketenagakerjaan di Kotamadya Ambon Propinsi Maluku (Youth, Education and Employment in the City of Ambon, Maluku)*, ed. Roosmalawati et al., 50–73. Jakarta: Puslitbang Kependudukan dan Ketenagakerjaan LIPI.

Djohansyah, Marliani. 1986. *Profil Kependudukan Kalimantan Selatan (Population Profile of Kalimantan Selatan)*. Jakarta: Kerjasama Meneg Kependudukan dan Lingkungan Hidup dan LDUI.

Dove, Michael R. (1985). Government Perceptions of Traditional Social Forestry in Indonesia: The History, Causes and Implications of State Policy on Swidden Agriculture. Unpublished paper.

Effendi, Tadjuddin Noer. 1997. Lintas Batas Internasional Pekerja: Perspektif Ketenagakerjaan (Cross-Border Migrant Workers: A Labor Perspective). Paper presented at *Simposium Mempersiapkan Instrumen Perlindungan untuk Buruh Migran Indonesia*, KOPBUMI, Solo, 30–31 Juli.

ELSAM. 1995. *Ke Arah Ratifikasi Konvensi Anti Penyiksaan: Kajian Kasus-kasus Penyiksaan Belum Terselesaikan (Towards a Ratification of Convention on Anti Torture: Study on the Unfinished Cases of Torture)*. Jakarta: Lembaga Studi dan Advokasi Masyarakat.

Elson, Robert E. 2001. *Suharto: A Political Biography*. Cambridge: Cambridge University Press.

Emmerson, Donald K., ed. 1999. *Indonesia Beyond Suharto: Polity, Economy, Society, Transition*. New York: M.E. Sharpe.

Environment and Settlement Studies Centre. 1979. *Circular Migrant in Jakarta*. Jakarta: Environment and Settlement Studies Centre.

ESCAP. 1981. *Migration, Urbanization and Development in Indonesia*. New York: United Nations, Country Report.

Esman, M.J. 1994. *Ethnic Politics*. Ithaca and London: Cornell University Press.

Evers, Hans-Dieter, Wolfgang Clauss, and Solvay Gerke. 1988. Population Dynamics, Ethnic Relations and Trade Among Javanese Transmigrants in East Kalimantan. Research Report Series No. 48. Population Studies Center Gajah Mada University.

Feith, Herbeth. 1980. Repressive-Developmentalist Regimes in Asia: Old Strengths, New Vulnerabilities. *Prisma* 19: 39–55.

Feith, H., and D. Lev. 1963. The End of the Indonesian Rebellion. *Pacific Affairs* 36 (1): 32–46.

Ferguson, R. Brian, ed. 2004. *The State Under Siege: Political Disintegration in the Post-Cold War Era*. New York: Gordon and Breach.

Finkle, Jason L., and C. Alison McIntosh, eds. 1994. *The Politics of Population: Conflict and Consensus in Family Planning. A supplement to Vol. 20*, Population and Development Review. New York: The Population Council.

Fitrani, Fitria, Bedrt Hofman, and Kai Kaiser. 2005. Unity in Diversity? The Creation of New Local Governments in a Decentralising Indonesia. *Bulletin of Indonesian Economic Studies* 41 (1): 57–79.

Forbes, Dean K. 1981. Mobility and Uneven Development in Indonesia. In *Population Mobility and Development: Southeast Asia and the Pacific. Development Studies Centre, Monograph no. 27*, ed. G.W. Jones and H.V. Richter. Canberra: Australian National University.

Frederick, W.H., and Robert L. Worden. 1993. *Indonesia: A Country Study*. Washington, DC: Federal Research Division, Library of Congress.

Fulcher, Mary Beth. 1983. Resettlement and Replacement: Social Dynamics in East Kalimantan, Indonesia. Unpublished Ph.D. thesis, Northwestern University.

Geertz, C. 1960. *The Religion of Java*. Glencoe: The Free Press.

Geertz, Clifford. 1963. *Agricultural Involution: The Processes of Ecological Change in Indonesia*. Berkeley: University of California Press.

———. 1976. *The Religion of Java*. Chicago: The University of Chicago Press.

Glassburner, B. 1978. Political Economy and the Suharto Regime. *Bulletin of Indonesian Economic Studies* 14 (3): 24–51.

Gondowarsito, R. 1986. Transmigrasi Bedol Desa: An Empirical Case Study of Inter-island Village Resettlement in Indonesia. Unpublished Ph.D. thesis, University of Bielefeld.

Grillo, R.D. 1998. *Pluralism and the Politics of Difference*. Oxford: The Clarendon Press.

Grindle, M., ed. 1980. *Politics and Policy Implementation in the Third World*. Princeton: Princeton University Press.

Guinness, Patrick, ed. 1977. *Transmigrants in South Kalimantan and South Sulawesi: Inter-island Government Sponsored Migration in Indonesia*, Report Series No. 15. Yogyakarta: Population Institute, Gajah Mada University.

Gunawan, I. 1986. Sebaiknya, Transmigrasi Tidak Bicara Target pada tahun 1986 (It would be better if Transmigration did not mention a target for 1986), Kompas, 8 January, 4–5.

Gurr, Ted Robert. 1993. *Minorities at Risk: A Global View of Ethnopolitical Conflicts*. Washington, DC: United States Institute of Peace Press.

———. 2000. *Peoples Versus States: Minorities at Risk in the New Century*. Washington, DC: United States Institute of Peace Press.

Hadiz, Vedi R. 1997. *Workers and the State in New Order Indonesia*. London: Routledge.

Hamidy, U.U. 1981. *Peranan Suku Banjar dalam Kehidupan Sosial Budaya di Kabupaten Inderagiri Hilir (The Impact of Banjarese People on the Social and Cultural Life in Inderagiri Hilir Regency)*. Pekanbaru: Lembaga penelitian Universitas Riau.

Hamidy, U.U., and M. Ahmad. 1984. *Masalah Sosial Budaya dan teknologi Transmigrasi Lokal (The Socio-Cultural and Technological Problems of Local Transmigration)*. Pekanbaru: Bappeda Riau.

Hamzah, U. 1980. Penyiapan Lahan Untuk Transmigrasi (The Land Preparation for Transmigration). Kompas, 2 June, pp. 5.

Handajaningrat, S., and R. Hindratmo. 1986. Landasan dan Pedoman Kerja Administrasi Pemerintah Daerah, Kota dan Desa (The Basic Guidelines for Regional, City and Village Government Administration).

Hanson, A.J. 1981. Transmigration and Marginal Land Development. In *Agricultural and Rural Development in Indonesia*, ed. G.E. Hansen. Boulder, CO: Westview Press.

Hanson, A.J., and Santo Koesoebiono. 1977. *Settling Coastal Swamplands in Sumatra: A Case Study for Integrated Resource Management*. Bogor: Center for Natural Resource Management and Environmental Studies, Bogor Agricultural University.

Hardjono, J.M. 1986. Transmigration: Looking to the Future. *Bulletin of Indonesian Economic Studies* 22 (2): 28–53.

Hardjosudarmo, S. 1965. *Kebijaksanaan Transmigrasi dalam rangka Pem-bangunan Masyarakat Desa di Indonesia (Transmigration Policy in the Context of Rural Development in Indonesia)*. Jakarta: Bhratara.

Hatta, Mohammad. 1954. *Beberapa Fasal ekonomi: Djalan ke Ekonomi dan Koperasi (Several Economic Aspects: Path Towards Economy and Cooperatives)*. Jakarta: Perpustakaan Perguruan Kementrian P.P. dan K.

Hayes, A. 1990. Population Quality and Sustainable Development in Indonesia. *Majalah Demografi Indonesia* XVII (34): 1–20.

Heeren, H.J. 1979. *Transmigrasi di Indonesia (Transmigration in Indonesia)*. Yogyakarta: Gajah Mada University Press.

Herrin, Alejandro. 1993. Population and Sustainable Development Issues. Paper prepared for *Seminar on Development Implications of Population Trends in Asia* sponsored by Australian National University, Canberra, Australia, September 29–October 1, 1993.

Hill, Hal. 1994. *Indonesia's New Order*. Sydney: Allen and Unwin.

———. 1996. *The Indonesian Economy Since 1966: Southeast Asia's Emerging Giant*. Cambridge: Cambridge University Press.

Homer-Dixon, Thomas. 1994a. Environmental Scarcities and Violent Conflict: Evidence from Cases. *International Security* 19 (1): 5–44.

———. 1994b. Population and Conflict. IUSSP distinguish lecture series on population and development, prepared in conjunction for the 1994 UN International Conference on Population and Development in Cairo.

Honggominarso, Suko B. 1977. Migration into the City of Medan, Indonesia: An Exploration of Migrant Adjustment. Unpublished MA Thesis, College of Social Sciences, Florida State University, Tallahassee.

Hooper, M.A.C. 1987. An Interview with the President of the Republic of Indonesia, Soeharto. *Kaleidoscope International* 9 (1): 25–28.

Hopper, M.A.C. 1987. Transmigration: Taming the Wilderness. *Kaleidoscope International* 9 (1): 194–197.

Hughey, Michael W., ed. 1998. *New Tribalism's: The Resurgence of Race and Ethnicity*. London: MacMillan Press Ltd.

Hugo, Graeme. 1978. *Population Mobility in West Java*. Yogyakarta: Gadjah Mada University Press.

———. 1980. Population Movements in Indonesia During the Colonial Period. In *Indonesia: The Making of a Culture*, ed. J.J. Fox. Canberra: Research School of Pacific Studies, The Australian National University.

———. 1982. Circular Migration in Indonesia. *Population and Development Review* 8 (1): 59–83.

———. 1993a. International Labor Migration and Family: Some Observations from Indonesia. *Asian and Pacific Migration Journal* 4 (2-3): 273–302.

———. 1993b. Indonesian Labor Migration to Malaysia: Trend and Policy Implications. *Southeast Asian Journal of Social Science* 21 (1): 36–70.

———. 1995. Labor Export from Indonesia: An Overview. In *Labor Migration in Asia (Special Focus). ASEAN Economic Bulletin*, ed. Philip L. Martin et al., vol. 12/2, 275–298. Singapore: Institute of Southeast Asian Studies.

———. 1997. Changing Patterns and Processes in Population Mobility. In *Indonesia Assessment: Population and Human Resources*, ed. Gavin W. Jones and Terrence H. Hull, 68–100. Canberra: Research School of Pacific and Asian Studies.

Hugo, Graeme J., T.H. Hull, V.J. Hull, and G.W. Jones. 1987. *The Demographic Dimension in Indonesian Development*. Singapore: Oxford University Press.

Hulme, D. 1987. State-Sponsored Land Settlement Policies: Theory and Practice. *Development and Change* 18: 413–436.

Human Right Watch. 1997. Indonesia: Communal Violence in West Kalimantan. *Report* 9(10).

Ignas Kleden's column. 2005. *Tempo* 13: 56–57.

Institut Pertanian Bogor dan Departemen Tenaga Kerja. 1997. *Evaluasi Dampak Penempatan Tenaga Kerja ke Luar Negeri*. Jakarta: Laporan Penelitian.

Jaspan, M.A. 1960. Persoalan Transmigrasi Obat Mudjarab atau Djalan buntu (Transmigration: A Panacea or A Dead End?), Karya Tak Berkala, No. 4. In *Lembaga Penyelidikan Kemasyarakatan*. Bandung: Universitas Padjadjaran.

Jellinek, L. 1977. The Pondok of Jakarta. *Bulletin of Indonesian Economic Studies* 13 (3): 67–71.

Jones, Gavin W. 1978. *The Life of the Poor in Indonesian Cities*, Annual Indonesia Lecture Series. Clayton: Monash University.

———. 1993. Population, Environment and Sustainable Development in Indonesia. *Majalah Demografi Indonesia* XX (40): 1–20.

Kahin, George McT. 1963. Indonesia. In *Major Governments of Asia*, ed. George McT. Kahin. Ithaca, NY: Cornell University Press.

Kantor Statistik Kalimantan Timur. 1998. Kalimantan Timur Dalam Angka.

Kantor Statistik Riau. 1986. Riau dalam angka.

Kanwil Transmigrasi Kalimantan Selatan. 1987. Program penyelenggaraan Transmigrasi dal am Pelita TV dan Kerangka Persiapan Rencana Program Penempatan Transmigrasi dalam Pelita V.

Kaplan, Robert D. 1994. The Coming Anarchy. *Atlantic Monthly*, February.

Kartodirdjo, Sartono. 1973. *Protest Movement in Rural Java: A Study of Agrarian Unrest in the Nineteenth and Early Twentieth Centuries*. Singapore: Oxford University Press.

Kassim, Azizah. 1993. The Registered and The Illegal: Indonesian Immigrants in Malaysia. Paper presented at *Seminar on Movement of Peoples in Southeast Asia*, organized by Southeast Asian Studies Center, LIPI, Jakarta 17–19 Mei.

———. 2009. Filipino Refugees in Sabah: State Responses, Public Stereotypes and the Dillema over Their Future. *Southeast Asian Studies* 47 (1): 52–88.

Kato, T. 1981. *Matriliny and Migration*. Ithaca: Cornell University Press.

King, Victor T. 1993. *The People of Borneo*. Oxford: Blackwell.

King, Russell, and Ronald Skeldon. 2010. Minds the Gap! Integrating Approaches to Internal and International Migration. *Journal of Ethnic and Migration Studies* 36 (10): 1619–1646.

Kingston, J. 1990. Agricultural Involution Among Lampung's Javanese? *Southeast Asian Studies* 27 (4): 485–507.

Kleden, Ninuk I. 1999. 'Metode Pemahaman Bagi Penelitian Antropologi', makalah tidak diterbitkan.

Kling, Zainal. 1985. The Socio-Cultural Unity of the Malay World. Paper presented at the Second Malay World Symposium, Colombo, Sri Lanka, 6 August.

Koentjaraningrat. 1975. Population Mobility in Villages Around Jakarta. *Bulletin of Indonesian Economic Studies* 11 (2): 108–119.

Kuntjorojakti, Dorodjatun. 1978. The Political Economy of Development: The Case Study of Indonesia Under the New Order Government. Unpublished Ph.D. dissertation, University of California, Berkeley.

Kurosawa, Aiko. 1993. *Mobilisasi dan Kontrol: Studi Tentang Perubahan Sosial di Pedesaan Jawa. 1942–1945*. Jakarta: Grasindo.

Leftwich, Adrian. 1983. *Redefining Politics: People, Resources and Power*. London and New York: Methuen.

Legge, John D. 1961. *Central Authority and Regional Autonomy in Indonesia: A Study in Local Administration 1950–1960*. Ithaca, NY: Cornell University Press.

———. 1964. *Indonesia*. Englewood Cliffs, NJ: Prentice Hall.

Levang, Patrice. 1995. Transmigration as a Medley of World's Resettlement Policies. Paper presented at the *International Seminar on Population Resettlement for Poverty Alleviation*, organized by Ministry of Transmigration and Forrest Squatter Resettlement, Jakarta 27 November–1 December.

Lim, L.L., and N. Oishi. 1996. International Labor Migration of Asian Women: Distinctive Characteristics and Policy Concerns. *Asian and Pacific Migration Journal* 5 (1): 85–116.

Lindbald, J.T. 2002. The Outer Islands in the 19th Century: Contest for the Periphery. In *The Emergence of a National Economy: An Economic History of Indonesia, 1800–2000*, ASAA Southeast Asian Publication Series, ed. Howard Dick et al. Crows Nest, NSW: Allen & Unwin.

Lineton, J. 1975a. Pasompe Ugi: Bugis Migrants and Wanderers. *Archipel* 10: 173–201.

———. 1975b. An Indonesian Society and Its Universe: A Study of the Bugis of South Sulawesi (Celebes) and Their Role Within a Wider Social and Economic System. Unpublished Ph.D. thesis, School of Oriental and African Studies, University of London.

Linquist, Johan. 2010. Labour Recruitment, Circuits of Capital, and Gendered Mobility: Reconceptualizing the Indonesian Migration Industry. *Pacific Affairs* 83 (1): 115–132.

Loveard, K. 1999. *Suharto, Indonesia's Last Sultan*. Singapore: Horizon Books.

Ludden, David. 2003. Presidential Address: Maps in the Mind and the Mobility of Asia. *The Journal of Asian Studies* 62 (4): 1057–1078.

MacAndrews, C., ed. 1986. *Central Government and Local Development in Indonesia*. Singapore: Oxford University Press.

MacAndrews, C., H.B. Fisher, and A. Sibero. 1982. Regional Development, Planning and Implementation in Indonesia: The Evolution of a National Policy. In *Too Rapid Rural Development: Perceptions and Perspectives from Southeast Asia*, ed. C. MacAndrews and C.L. Sien, 79–121. Athens: Ohio University Press.

Mackie, J.A.C. 1974. *Konfrontasi: The Indonesia-Malaysia Dispute 1963–1966*. Kuala Lumpur: Oxford University Press.

———. 1980. Integrating and Centrifugal Factors in Indonesia's Politics Since 1945. In *Indonesia: The Making of a Nation, Vol. 2 of Indonesia: Australian Perspectives*, 669–684. Canberra: Research School of Pacific Studies, Australian National University.

Magenda, Burhan D. 1986. Ethnicity and State-Building in Indonesia: The Cultural Base of the New Order. In *Ethnicities and Nations: Processes of Interethnic Relations in Latin America, Southeast Asia, and the Pacific*, ed. Remo Guideri et al. Austin: A Rothko Chapel Book, University of Texas.

Magenda, Burhan. 1989. The Surviving Aristocracy in Indonesia: Politics in Three Provinces of the Outer Islands (Volume I and II). Unpublished Ph.D. thesis, Cornell University.

Mangkusuwondo, S. 1970. Repelita: The First Six Month. In *Five Papers on Indonesian Economic Development*, 15–31. Jakarta: Yayasan Badan Penerbit Fakultas Ekonomi Universitas Indonesia.

Manning, Chris, and Michael Rumbiak. 1989. Irian Jaya: Economic Change, Migrants, Indigenous Welfare. In *Unity and Diversity: Regional Economic Development Since 1970*, ed. Hal Hill, 77–106. Singapore: Oxford University Press.

Mantra, I.B. 1981. *Population Movement in Wet Rice Communities: A Case Study of Two Dukuh in Yogyakarta Special Region*. Yogyakarta: Gadjah Mada University Press.

Mantra, Ida Bagoes. 1992. Pola dan Arah Migrasi Penduduk Antar Propinsi di Indonesia Tahun 1990 (Pattern and Trend of Population Movement in Indonesia, 1990). *Populasi* 2 (3): 24–38.

Martin, Philip L., Andrew Mason, and Ching-lung Tsay, eds. 1995. *Labor Migration in Asia. ASEAN Economic Bulletin (Special Focus)*. Vol. 12/12. Singapore: Institute of Southeast Asian Studies.

Maryanov, G.S. 1958. *Decentralization in Indonesia as a Political Problem*. Ithaca, NY: Modern Indonesia Project, Cornell University.

Mas'oed, M. 1983. The Indonesian Economy and Political Structure During the Early New Order, 1966–71. Unpublished Ph.D. dissertation, Ohio State University.

Maybury-Lewis, David. 1997. *Indigenous Peoples, Ethnic Group and the State. Cultural Survival Studies in Ethnicity and Change*. Boston: Allyn and Bacon.

McCawley, Peter. 1981. The Growth of the Industrial Sector. In *Indonesian Economy During the Suharto Era*, ed. Anne Booth and Peter McCawley, 62–101. Kuala Lumpur: Oxford University Press.

McCutheon, L. 1978. 'Occupation' and Housing Adjustment of Migrants of Surabaya, Indonesia: The Case of Second City. *International Migration Review* 12 (1): 82–92.

McNicoll, Geoffrey. 1968. Internal Migration in Indonesia: Descriptive Notes. *Indonesia* 5: 29–92.

McVey, Ruth. 1996. Building Behemoth: Indonesian Construction of the Nation-State. In *Making Indonesia: Essays on Modern Indonesia in Honor of George Mc T. Kahin*, ed. Daniel S. Lev and Ruth McVey. Ithaca, NY: Cornell Southeast Asia Program.

Meneg Kependudukan/BKKBN. 1997. Mobilitas Penduduk dan Pembangunan Daerah: Analisis SUPAS 95. In *Laporan Hasil Kerjasama Kantor Menteri Negara Kependudukan/BKKBN dan Lembaga Demografi FEUI*. Jakarta: Meneg Kependudukan/BKKBN.

Meyer, P.A., and C. MacAndrews. 1978. *Transmigration in Indonesia: An Annotated Bibliography*. Jogjakarta: Gadjah Mada University Press.

Ministry of Transmigration and Forest Squatter Settlement. 1993. *Planning for Forest Squatter Resettlement: A Draft Report*. Jakarta: Ministry of Transmigration and Forest Squatter Settlement.

Mink, Stephen D. 1992. Poverty, Population, and the Environment. World Bank Discussion Papers. Washington, DC: The World Bank.

Miyazaki, Koji. 1997. Javanese-Malay: Colonial Immigrants in Post-Colonial Era. Paper presented at the workshop on *Migration in Contemporary Southeast Asia*, ISEAS, Singapore, 22–23 January.

———. 1998. Culture Moves: Contemporary Migration in Southeast Asia. Keynote Speech (2), at the symposium on Population Movement in Southeast Asia: Changing Identities and Strategies for Survival, September 17–19, at National Museum of Ethnology, Osaka, Japan.

Montgomery, R.D. 1975. Migration, Employment and Unemployment in Java: Changes from 1961–1971, with Particular Reference to the Green Revolution. *Asian Survey* 15 (3): 224–234.

Morfit, Michael. 1984. Strengthening the Capacities of Local Government in Indonesia: Issues, Strategies, and Experiences. *Contemporary Southeast Asia* 6 (1).

Muhidin, Salut. 2021. Demografi dan Demokrasi di Indonesia setelah Masa Reformasi (Demography and Democracy in Indonesia's post Reformation). Paper presented at the Webinar on Social and Cultural Transformation in the post-reformation Indonesia organized by The Center for Social and Cultural Studies Indonesian Institute of Science (PMB-LIPI) and The Center for Heritage Nusantara Studies Christian University of Satya Wacana (PSHN-UKSW), 16 February.

Murad, Auda. 1980. *Merantau: Outmigration in a Matrilineal Society of West Sumatra*, Indonesian Population Monograph Series, No. 3. Canaberra: Department of Demography, Australian National University.

Naim, Mochtar. 1979. *Merantau: Pola Migrasi Suku Minangkabau (Merantau: The Migration Pattern of Minangkabau)*. Jogjakarta: Gadjah Mada University Press.

Nawawi, M.A. 1968. Regionalism and Regional Conflicts in Indonesia. Unpublished Ph.D. thesis, Princeton University, Princeton.

Nitisastro, W., and Nathan Keyfitz. 1955. *Soal Penduduk dan Pembangunan (Population and Development)*. Jakarta: PT Pembangunan.

Nitisastro, Widjojo. 1970. *Population Trends in Indonesia*. New York: Cornell University Press.

Oey, Mayling. 1977. Jakarta Dibangun Pendatang (Jakarta Builds by Migrants). *Prisma* 5.

———. 1981. The Impact of Migration on Fertility: A Case Study of Transmigrants in Lampung, Indonesia. Unpublished Ph.D. Thesis, Department of Demography, ANU, Canberra.

ORSTROM dan Departemen Transmigrasi. 1986. Transmigration and Spontaneous Migrations, Provinsi Lampung Indonesia.

Osborne, R. 1985. Transmigration: The Great Population Debate Hots-Up. *Inside Indonesia* 6: 22–26.

Paauw, D.S. 1963. From Colonial to Guided Economy. In *Indonesia*, ed. Ruth McVey. New Haven: Southeast Asia Studies, Yale University.

Pangestu, Mari. 1991. East Kalimantan: Beyond the Timber and Oil Boom. In *Unity and Diversity: Regional Economic Development in Indonesia Since 1970*, ed. Hal Hill, 151–176. Singapore: Oxford University Press.

Pangestu, Mari, and Iwan J. Aziz. 1994. Survey of Recent Development. *Bulletin of Indonesian Economic Studies* 30 (2): 3–47.

Papanek, G.F. 1975. The Poor of Jakarta. *Economic Development and Cultural Change* 24 (1): 1–27.

Parta, Made Ngurah. 1996. Karakteristik Pedagagang Kaki Lima Suku Bugis: Studi Kasus di Pasar Citra Niaga Kotamadya Samarinda. Master thesis at Padjadjaran University Graduate Program.

Partadiredja, A. 1970. Economic Survey of South Kalimantan. *Bulletin of Indonesian Economic Studies* VI (2) 46–65. July.

Pelly, Usman. 1983. Urban Migration and Adaptation in Indonesia: A Case Study of Minangkabau and Mandailing Batak Migrants in Medan, North Sumatra. Unpublished Ph.D. Thesis, Graduate College of the University of Illinois, Urbana-Champaign.

———. 1994. *Urbanisasi dan Adaptasi: Peranan Misi Budaya Minang-kabau dan Mandailing*. Jakarta: LP3ES.

Pelras, Christian. 1994. *The Bugis*. Oxford: Blackwell.

Peluso, Nancy Lee. 1987. Merchant, manipulation, and minor forest product on the Mahakam: Bugis political economic strategy in pre-colonial Kutai. International Workshop on Indonesian Studies No. 2, Royal Institute of Linguistics and Anthropology, Leiden, 2–6 November.

———. 1998. Rich Forest, Poor People and Development: Forest Access Control and Resistance in Java. Unpublished Ph.D. dissertation, Cornel University.

Pelzer, Krl J. 1945a. *Pioneer Settlement in the Asiatic Tropics*. New York: Institute of Pacific Relations.

Pelzer, K.J. 1945b. *Pioneer Settlement in the Asiatic Tropics*. American Geographical Society, Special Publication No. 29. New Haven: Yale University Press.

Pemberton, J. 1994. *On the Subject of Java*. Ithaca and London: Cornell University Press.
Penders, C.L.M. 1969. Java's Population Problem During the Colonial Period. *World Review* 8 (1): 24–34.
———., ed. 1977. *Indonesia: Selected Documents on Colonialism and Nationalism, 1930–1942*. St. Lucia: University of Queensland Press.
PEP-LIPI. 1997. *Otonomi Daerah: Laporan Kegiatan Pengembangan Ilmu-Ilmu Sosial (Regional Autonomy: Report on Social Sciences Development Activities)*. Jakarta: Puslitbang Ekonomi dan Pembangunan, Lembaga Ilmu Pengetahuan Indonesia.
Perez-Sainz, J.P. 1983. Transmigration and Accumulation in Indonesia. In *State Policies and Internal Migration*, ed. A.S. Oberai, 183–228. London: Croom Helm and St. Martin Press.
Pillai, Patrick. 1992. *People on the Move: An Overview of Recent Immigration and Emigration in Malaysia, Research Paper*. Kuala Lumpur: Institute of Strategic and International Studies (ISIS).
———. 1995. Malaysia. In *Labor Migration in Asia. ASEAN Economic Bulletin (Special Focus)*, ed. Philip L. Martin et al., vol. 12/12, 221–236. Singapore: Institute of Southeast Asian Studies.
Potter, Lesley. 1993. Banjarese in and Beyond Hulu-Sungai, South Kalimantan: A Study of Cultural Dependence, Economic Opportunity and Mobility. In *New Challenges in Modern Economic History of Indonesia*, ed. J.Th. Lindblad, 264–298. Leiden: Program of Indonesian Studies.
Potter, L., and A. Hasymi. 1987. Indonesian Regional Economic Surveys: South Kalimantan, The Banjarese Heartland. Paper presented at the Regional economic Surveys Workshops, February, Australian National University.
PPT-LIPI. 1994. *Mobilitas Penduduk Perbatasan Di Irian Jaya (Population Mobility in the Border Area of Irian Jaya)*. Jakarta: Puslitbang Kependudukan dan Ketenagakerjaan, LIPI.
Primawati, Anggraeni. 2008. Migrasi internasional dan perubahan masyarakat lokal: suatu studi mengenai proses dan dampak mobilitas warga Kecamatan Purwodadi, Kabupaten Purworejo, Provinsi Jawa Tengah, ke Malaysia. Unpublished PhD thesis, Department of Sociology, University of Indonesia, Jakarta.
Putra, Ahimsa, and Heddy Shri. 1989. *Minawang: Hubungan Patron-Klien di Sulawesi Selatan (Minawang: Patron-Client Relation in South Sulawesi)*. Yogyakarta: Gajah Mada University Press.
Quick, S.A. 1980. The Paradox of Popularity: Ideological Program Implementation in Zambia. In *Politics and Policy Implementation in the Third World*, ed. M. Gridle, 40–63. Princeton: Princeton University Press.
Raharto, Aswatini. 1997. Going *Melarat* for Better Living. *Eastern Indonesia Population and Development Research Project Newsletter* 3 (3): 3–5.
———, ed. 1996. *Mobilitas Penduduk Wilayah Perbatasan Indonesia-Malaysia di Kalimantan Timur*. Jakarta: Puslitbang Kependudukan dan Ketenagakerjaan, LIPI.
Raharto, Aswatini, and Haning Romdiati. 1998. Mobilitas Penduduk dan Pembangunan di KTI (Population Mobility and Development in Eastern Indonesia). Paper presented at the LIPI Seminar, Manado, 18–19 February 1998.
Ramadhan, K.H., and G. Dwipayana. 1989. *Soeharto: Pikiran, Ucapan dan Tindakan Saya*. Jakarta: PT. Citra Lamtoro Gung Persada.
Redecon (Resource Development Consultants). 1986. Laporan Hasil Studi Kebijaksanaan Pengelolaan Program Transmigrasi (A report on the study of policy and management of Transmigration Program), Jakarta.
Ricklefs, M.C. 1981. *A History of Modern Indonesia*. London: MacMillan.
Robison, Richard, and Vedi R. Hadiz. 2004. *Reorganising Power in Indonesia: The politics of Oligarchy in an Age of Market*. London and New York: Routledge Curzon.
Rodolfo, Stavenhagen, ed. 1996. *Ethnic Conflicts and the Nation-State*. London: MacMillan Press.
Rosadi, Imron. 2010. Konsekuensi migrasi internasional terhadap relasi gender: studi tentang buruh migran internasional yang telah pulang kembali ke keluarganya di Kecamatan Juntinyuat,

Kabupaten Indramayu, Provinsi Jawa Barat. Unpublished PhD thesis, Department of Sociology, University of Indonesia, Jakarta.
Rostow, Walter W. 1960. *The Stages of Economic Growth*. Cambridge: Cambridge University Press.
Rudner, M. 1976. The Indonesian Military and Economic Policy: The Goals and Performance of the First Five-Years Development Plan, 1969–1974. *Modern Asian Studies* 10: 244–284.
Rumbiak, Michael C. 1984. Nimboran Migration to Jayapura Irian Jaya and Rural-Urban Ties. Unpublished M.A. thesis, Department of Demography, the Australian National University.
Sadli, M. 1963. Indonesia's Hundred Millions. *Far Eastern Economic Review* 42: 21–23.
———. 2001. Nurturing Optimism About the Future. *Jakarta Post,* May 29.
Saleh, M.I. 1962. *Banjarmasin*. Jakarta: Balai Pendidikan Guru.
Salim, Emil. 1970. Satu Tahun Repelita (The First Year of the First Five-Year Development Plan). Paper presented at the Indonesian Economic Association (ISEI) Annual Conference, 17 March, Yogyakarta.
Samego, Indria. 1996. Mewujudkan Mimpi Bersama: Beberapa Catatan dari Pertemuan Usahawan Serantau. *Republika,* April 30.
Santoso, A. 1986. Public Policy Implementation: Rice Policy at the Regional Level in Indonesia 1970–1984. Unpublished Ph.D. thesis, Australian National University, Canberra.
Santoso, Djoko, and Ali Wardhana. 1957. Beberapa Segi Transmigrasi Spontan di Indonesia (Some Aspects of Spontaneous Transmigration in Indonesia). *Ekonomi dan Keuangan Indonesia* 2: 98–115.
Schierup-Carl, Ulrik, Peo Hansen, and Stephen Castles. 2006. *Migration, Citizenship, and European Welfare State*. Oxford: Oxford University Press.
Schultz, L. 1972. Politics of Development Planning in Indonesia. Unpublished Ph.D. thesis, Claremont Graduate School, Claremont.
Sebastian, Leonard C. 1996. Indonesian National Security and Defense Planning. Unpublished Ph. D. thesis, Australian National University.
———. 2006. *Real Politic Ideology: Indonesia's Use of Military Force*. Singapore: Institute of Southeast Asian Studies.
Sethuraman, S.V. 1976. *Urbanization and Employment in Jakarta*. Geneva: International Labour Organization.
Setiadi. 2010. Dinamika otonomi perempuan desa: studi dampak migrasi internasional di pedesaan Jawa. Unpublished PhD thesis, Department of Anthropology, University of Gajah Mada, Yogyakarta.
Shah, N.M., and Indu Menon. 1997. Violence Against Women Migrant Workers: Issues, Data and Partial Solutions. *Asian and Pacific Migration Journal* 6 (1): 5–30.
Shiraishi, T. 1981. The Disputes Between Tjipto Mangoenkoesoemo and Soetatmo Soeriokoesomo: Satria vs. Pandita. *Indonesia* 32: 93–108.
Sinaga, R.S. 1978. Implications of Agricultural Mechanization for Employment and Income Distribution: A Case Study from Indramayu, West Java. *Bulletin of Indonesian Economic Studies* 14 (2): 102–111.
Singarimbun, M., and D. Penny. 1976. *Penduduk dan Kemiskinan: Kasus Sriharjo di Pedesaan Jawa (Population and Poverty: The Case of Sriharjo in Rural Java)*. Jakarta: Bhatara Aksara.
Sjahrir. 1987. *Kebijaksanaan Negara: Konsistensi dan Implementasi (Public Policy: Consistency and Implementation)*. Jakarta: LP3ES.
Sjamsu, Amral. 1960. *Dari Kolonisasi ke Transmigrasi, 1905–1955 (From Colonization to Transmigration, 1905–1955)*. Jakarta: Penerbit Djambatan.
Smith, Thomas, 1987. Unpublished manuscript.
Soetrisno, Loekman, et al. 1991. *East Timor: The Impact of Integration. An Indonesian Socio-Anthropological Study*. Northcote: IRIP.
Soewartoyo. 1994. Migrasi Internal di Timor Timur: Kajian di daerah Tujuan pasca-Integrasi (Internal Migration in East Timor: A study on Destination Area—Post Integration). *Analisis* XXVI (3): 265–275.

Spaan, E. 1994. Taikongs and Calos: The Role of Middlemen and Brokers in Javanese International Migration. *International Migration Review* 28 (1): 93–119.

Speare, A. 1975. Interpreting the Migration Data from the 1971 Census (Indonesia). *Majalah Demografi Indonesia* 3: 66–85.

Steele, Ross McLean. 1980. Origins and Occupational Mobility of Life-time Migrant to Surabaya, East Java. Unpublished Ph.D. thesis, Department of Geography, The Australian National University.

Subroto. 1974. Kebijaksanaan di Bidang Kesempatan Kerja dan Transmigrasi dalam Repelita II (Government Policies on Employment Opportunity and Transmigration in the Second Five-Year Development Plan). *Prisma* 2: 18–29.

Sugardo. 1986. *Pengaruh Distribusi Penduduk terhadap Geo-Politik dan Geo-Strategi Indonesia (The Impact of Population Distribution on Geo-politics and Geo-strategy)*. Jakarta: Kantor Menteri Negara Kependudukan dan Lingkungan Hidup.

Suharso. 1978. The Pattern of Population Movement and Urbanization in Java 1968–1973. Unpublished Ph.D. thesis. University of Gadjah Mada, Yogyakarta.

Suharso, A. Speare, H.R. Redmana, and I. Husin. 1976. *Rural-Urban Migration in Indonesia*. Jakarta: National Institute of Economic and Social Research.

Sukarnaprawira. 1975. *Kalimantan Barat Daerah Khatulistiwa*. Ganaco Press.

Sulistyo, Hermawan. 1995. Negara dan Masyarakat Lokal: Studi tentang Otonomi dan Kapasitas Masyarakat Lokal (State and Local Society: Study on Regional Autonomy, Local Government, and the Capacity of Local Society). In *Negara dan Kemiskinan di Daerah (State and Poverty in the Regions)*, ed. Didik J. Rachbini dkk. Jakarta: Pustaka Sinar Harapan.

Sundrum, R.M. 1976. Interprovincial Migration. *Bulletin of Indonesian Economic Studies* 12 (1): 66–85.

Suparlan, P. 1981. Kata Pengantar. Buku Terjemahan Karya Clifford Geertz, Abangan, Santri, Priyayi Dalam Masyarakat Jawa, Pustaka Jaya, Jakarta.

Suparlan, Parsudi. 1995. *The Javanese in Suriname: Ethnicity in an Ethnically Plural Society*, Monograph Series. Temple: Program for Southeast Asian Studies, Arizona State University.

Surapaty, C., et al. 1993. Demographic Pressure and Deforestration in Lahat District of South Sumatra, Indonesia. Paper contributed to 22nd International Population Conference International Union for Scientific Study of Population, Montreal, August.

Suryadinata, L., et al. 2003. *Indonesia's Population: Ethnicity and Religion in a Changing Political Landscape*. Singapore: ISEAS.

Swain, Ashok. 1996. Environmental Migration and Conflict Dynamics: Focus on Developing Regions. *Third World Quarterly* 17 (5): 959–973.

Tadjoeddin, M.Z. 2002. *Anatomy of Social Violence in the Context of Transition: The Case of Indonesia 1990–2001*. Jakarta: United Nations Support Facility for Indonesian Recovery UNSFIR.

TAD-Report No.17. 1980. *Regional Planning for East Kalimantan*. East Kalimantan Transmigration Area Development Project PTA 46/ATA 140, TAD-*Bappeda* .

Tanter, Richard. 1991. Intelligence Agencies and Third World Militarization: A Case Study of Indonesia, 1969–1989. Unpublished Ph.D. thesis, Department of Politics, Monash University, Australia.

Teitelbaum, Michael S., and Jay Winter. 1998. *A Question of Numbers: High Migration, Low Fertility, and the Politics of National Identity*. New York: Hill & Wang.

Temple, G. 1975. Migration to Jakarta. *Bulletin of Indonesian Economic Studies* 11 (1): 76–81.

The Siaw Giap. 1981. Rural Unrest in West Kalimantan: The Chinese Uprising in 1914. Paper presented at the conference held in celebration of the fiftieth anniversary of the Sinological Institute of Leyden University, December 8–10, 1980/ed. by W.L. Idema.

Thee, Kian Wie. 1969. Plantation Agriculture and Export Growth: An Economic History of East Sumatra, 1963–1942. Unpublished Ph.D. dissertation, University of Wisconsin.

Thompson, V. 1947. *Labor Problems in Southeast Asia*. New Haven: Yale University Press.

Thompson, G., and R.C. Manning. 1974. The World Bank in Indonesia. *Bulletin of Indonesian Economic Studies* 10 (2): 56–82.

Tirtosudarmo, Riwanto. 1984. Migration Decision Making: The Case of East Java. Unpublished M.A. thesis, Department of Demography, The Australian National University, Canberra.

———. 1990. Transmigration and Its Center-regional Contexts: The Case of Riau and South-Kalimantan Provinces. Unpublished Ph.D. thesis, Research School of Social Sciences, The Australian National University, Canberra, Australia.

———. 1994. *A Paradigm Shift: From Transmigration to Human Resources Migration*. Paper presented in Human Resource Session at the Fourth East Asian Economic Association Conference, 25–26 August 1994, Taipei, Taiwan.

———. 1995. The Human Resources Development in Eastern Indonesia. Paper presented at the Indonesia Update Conference, Research School of Pacific and Asian Studies, Australian National University, Canberra, Australia, 25–26 August.

———. 1996. *Dari Riau sampai Timor-Timur: Demografi-Politik Pembangunan Indonesia (From Riau to East Timor: The Political Demography of Development in Indonesia)*. Jakarta: Pustaka Sinar Harapan.

———. 1997. Migration, Economic Development and Ethnic Conflict in Indonesia: A Preliminary Observation. *Sojourn* 12 (2): 293–328.

———. 1998. Labor Movement and the Quest for Democracy: A Twisted Working Class in New Order's Indonesia? Paper presented at the workshop on Discourses and Practices of Democracy in Southeast Asia, organized by Indonesian Institute Sciences and Gajah Mada University, Yogyakarta, 16–20 March.

———. 1999. The Indonesian State's Response to Migration. *Sojourn* 14 (1): 212–228.

———. 2001. Demography and Security: The Transmigration Policy in Indonesia. In *Demography and National Security*, ed. Myron Weiner and Sharon Stanton, 199–227. Boston: Berghahn Books.

Tirtosudarmo, R. 2005. The Orang Melayu and Orang Jawa in the Land Below The Wind. *Working Paper No. 14*. CRISE Oxford University.

Tirtosudarmo, Riwanto. 2006. In the Margin of a Borderland: The Florenese Community Between Nunukan and Tawau. In *Centering the Margin: Agency and Narrative in Southeast Asian Borderlands*, ed. Alexander Horstmann and Reed L. Wadley, 135–154. New York and Oxford: Berghahn Books.

———. 2015. Managing population under Jokowi. *The Jakarta Post*, June 25.

———. 2018. *The Politics of Migration in Indonesia and Beyond*. Singapore: Springer and LIPI Press.

Tirtosudarmo, Riwanto, and Titik Handayani. 1993. Kesenjangan Pendidikan dan Kesempatan Kerja di Timor Timur (Education and Employment Gap in East Timor). *Analisis* 6: 476–497.

Tirtosudarmo, Riwanto, and Haning Romdiati. 1998. *A Needs Assessment Concerning Indonesian Women Migrant Workers to Saudi Arabia*. Jakarta: Centre for Population and Manpower Studies, Indonesian Institute of Sciences.

Titus, M.J. 1978. Inter-regional Migration in Indonesia as a Reflection of Social and Economic Activities. *Tijdschrift voor Economische en Sociale Geografie* 62 (4): 194–204.

Tjokroamidjojo, B. 1979. The Way Abung Transmigration Project: Indonesia. In *Management of Development Project: An International Case Study*, ed. L.J. Goodman and R.N. Love. New York: Pergamon Press.

Tjondronegoro, S.M.P. 1972. Land Reform or Land Settlement: Shifts in Indonesia's Land Policy. *Land Tenure Center Paper*, No. 81. Madison: University of Wisconsin.

———. 1984. Population Pressure and Socio-Economic Development in Rural Indonesia. *Kabar Seberang* 13/14: 62–81.

Todaro, Michael P. 1978. *Economic Development in the Third World*. New York: Longman.

Ulaen, Alex. 2011. Dua Nama Satu Juragan: Sejarah Wilayah Perbatasan Miangas-Filipina 1928–2010. Unpublished draft report for the Ministry of National Education.

Utomo, Kampto. 1975. *Masyarakat Transmigran Spontan di Daerah Wai Sekampung, Lampung (Spontaneous Transmigrants in Wai Sekampung, Lampung).* Yogyakarta: Gadjah Mada University Press.

Van den Ham, A. 1989. An Analysis of Strategic Development Framework in Indonesia. *Regional Development Dialogue* 10 (2): 218–227.

Van der Goot, P.F. 1974. The Role and Participation of Foreign Aid in Transmigration Projects in Indonesia. In *The Role of Transmigration in National and Regional Development.* Jakarta: Proceeding of the 1971 Transmigration Workshop.

Van der Kroef, J. 1956. Population Pressure and Economic Development in Indonesia. In *Demographic Analysis: Selected Readings*, ed. J.J. Spengler and O.D. Duncan, 739–754. Illinois: Glencoe.

Varshney, Ashutosh, Rizal Panggabean, and Mohammad Zulfan Tadjoeddin. 2004. Patterns of Collective Violence in Indonesia (1990–2003). *UNSFIR Working Paper-04/03.*

Vayda, Andrew P., and Ahmad Sahur. 1996. *Bugis Settlers in East Kalimantan's Kutai National Park: Their Past and Present and Some Possibilities for Their Future.* Jakarta: CIFOR Special Publication.

Vidyattama, Yogi. 2008. Patterns of Provincial Economic Growth in Indonesia. Unpublished PhD thesis, The Australian National University.

Visser, L. 1988. An Interview with Koentjaraningrat. *Current Anthropology* 29 (5): 749–753.

Vredenbreght, J. 1964. Bawean Migrations. *Bijdragen tot de Taal-, Land-en Volkenkunde* 120: 109–139.

Vredenbregt, J. 1964. Masyarakat Perantau Pulau Bawean di Singapura (The Baweanese in Singapore). Unpublished Ph.D. thesis.

Warren, James. 1981. *The Sulu Zone 1768–1898: The dynamics of external trade, slavery, and ethnicity in the transformation of a Southeast Asian Maritime State.* Singapore: Singapore University Press.

Wee, Vivienne. 1993. *Jemberang:* Crossing the Straits of Melaka, Singapore and Riau. Paper prepared for the conference on Movement of Peoples in Southeast Asia, organized by the LIPI, Jakarta, Indonesia, 17–19 February.

Weiner, Myron. 1971. Political Demography: An Inquiry into the Political Consequences of Population Change. In *National Academy of Sciences, Rapid Population Growth: Consequences and Policy Implications.* Baltimore: The John Hopkins Press.

Weiner, M. 1975. Internal Migration Policies: Purposes, Interests, Instruments, Effects. In *Policy Sciences and Population*, ed. W.F. Ilchman et al., 65–94. Massachusetts: Lexington Books.

Weiner, Myron. 1978. *Sons of the Soil: Migration and Ethnic Conflict in India.* Princeton: Princeton University Press.

———. 1995a. Security, Stability and International Migration. In *Global Dangers: Changing Dimensions of International Security*, ed. Sean M. Lynn-Jones and Steven E. Miller, 183–218. Cambridge: The MIT Press.

———. 1995b. *The Global Migration Crisis: Challenges to States and Human Rights.* New York: Harper Collins.

Wertheim, W.F. 1959. Sociological Aspects of Inter-islands in Indonesia. *Population Studies* 7 (3): 184–201.

———. 1984 The Forgotten Ones of Java: The Sociologists False Consciousness. *Kahar Seberang* (13/14) 1–16. December.

———. 1989. The Lampung Affairs: A Personal Perspective. *Inside Indonesia* 19: 20–21.

White, B. 1976. Population, Involution and Employment in Rural Java. *Development and Change* 7: 267–290.

Wibowo, I. 1983. Migration and Squatter Settlements in Indonesia: A Chi-Square Approach to Systemic Social Problems as a Basis for Policy Formulation. Unpublished Ph.D. thesis, Graduate College, Texas A&M University.

Widayatun. 1988. *Potensi Daerah Transmigrasi di Kalimantan Selatan (Regional Resources and Transmigration in South Kalimantan).* Jakarta: Laporan Hasil Penelitian.

Widjojo, Muridan S. 1995. Pemberdayaan Masyarakat 'Lain-Lain' di Timika Irian Jaya (Empowering "the Others" in Timika, Irian Jaya). *Analisis-CSIS* XXVI (3): 296–320.

Wirosuhardjo, K. 1980. Population Mobility: The Development Policy Issues in Indonesia. Paper presented to the conference on Population Mobility and Development. Canberra: ANU, 8–10 October.

———. 1984. Pattern and Trends of Internal Migration and Urbanization in Indonesia and Their Policy Applications. Unpublished Ph.D. thesis, University of Indonesia, Jakarta.

Wood, William Bruce. 1985. Intermediate Cities in the Resource Frontier: A Case Study of Samarinda and Balikpapan, East Kalimantan, Indonesia. Unpublished Ph.D. thesis, University of Hawaii.

World Bank. 1988. *Indonesia: The Transmigration Program in Perspective*. Washington, DC: The World Bank.

———. 1995. *World Development Report 1995: Workers in an Integrating World*. Oxford: World Bank.

Young, Crawford, ed. 1993. *The Rising Tide of Cultural Pluralism: The Nation State at Bay?* Madison: The University of Wisconsin Press.

Zakaria, Yando. 1993. *Hutan dan Kesejahteraan Masyarakat (Forest and People's Welfares)*. Jakarta: Wahana Lingkungan Hidup Indonesia.

Zelinsky, W. 1971. The Hypothesis of the Mobility Transition. *Geographical Review* 41 (2): 219–149.

Index

R. Tirtosudarmo, *From Colonization to Nation-State*,
https://doi.org/10.1007/978-981-16-6437-3

G

H

I

J

K

L